FOSSILS

THE KEY TO THE PAST

NEW EDITION

FOSSILS

THE KEY TO THE PAST

NEW EDITION

RICHARD FORTEY

HARVARD UNIVERSITY PRESS
CAMBRIDGE, MASSACHUSETTS
1991

Printed in Great Britain

10   9   8   7   6   5   4   3   2   1

Library of Congress Cataloging-in-Publication Data
Fortey, Richard A.
   Fossils: the key to the past/Richard Fortey. – New ed.
     p. cm.
   Includes index.
   ISBN 0-674-31135-3
   1. Fossils.   2. Paleontology.   I. Title.
QE711.2.F66   1991
560–dc20   90-24760
                               CIP

CONTENTS

For Dominic

PREFACE

In this book I have tried to show how fossils, far from being mere dry bones, can be used to reconstruct the history of the Earth. My concern is *not* to present a catalogue of the different kinds of fossils that lived at various times in the vast stretches of the geological past. Rather I am concerned with how and why this past is reconstructed, how fossils link into the exciting geological discoveries of the last few years, and with areas of controversy which are still under debate. Palaeontology, for all its concern with dead animals and plants, is a living science, and one which is exciting only so long as it is changing its perspectives on interpreting the past.

I have had to venture into areas of palaeontology in which I am not an expert; particularly over controversial issues I have been forced to be partial, where a more leisurely approach might have allowed an exhaustive list of qualifications. One of the problems of the panoramic view is that detail can often only be hinted at. But wherever possible I have discussed general principles with reference to the specific example. Geology is one of the sciences with rather more exceptions than rules. My intention has been to define the rules, without overriding the true complexity of past history. I have tried to be simple – by keeping technical jargon to a minimum, for example – but without, I hope, being too simplistic.

One of the real excitements of working with fossils is the discovery of beautiful and informative specimens. I have selected for colour illustration a wide variety of the kind of fossils that the amateur collector has a good chance of discovering. The reconstructions of dinosaurs, and illustrations of exceptional or microscopic fossils, are figured in the text. The humblest and most common fossils are often among the most beautiful, and more important in reconstructing a picture of the past than the most spectacular rarities.

I am indebted to several of my colleagues at the British Museum (Natural History) for their advice and guidance on many details. Dr Euan Clarkson read through the manuscript, and made a number of valuable suggestions for its improvement. Ray and Corinne Burrows ably drew the text figures. The skill of the photographers at the British Museum (Natural History) is displayed in the colour plates; particular thanks are due to Peter Green and Colin Keates for supplying additional black and white photographs. Other photographs in the text were supplied by the following, whose help is gratefully acknowledged: The Palaeontological Association, *Punch*, Geological Society of London, The National Museum of Wales, Institute of Geological Sciences; Professors W. Stürmer, H. B. Whittington, G. Kelling; Drs K. M. Towe, G. Playford, J. M. Pettitt, J. D. Taylor, and D. P. Wilson.

Lastly, I thank my wife Jacqueline for her tolerance and support during the many hours I spent in the company of a typewriter, and for reading the proofs.

PREFACE TO THE NEW EDITION

In preparing a second edition of this book I wished to bring it up to date by taking into account some of the important discoveries which have been made in the eight years since the publication of the first edition. These discoveries include the long awaited 'conodont animal' – which had been an enigma for more than a hundred years – and I have taken the exciting new dinosaur *Baryonyx* as a case history of the work involved in reconstructing extinct animals, and inferring how they might have lived. Much work has been done on extinctions, and even as I write new ideas are appearing in the scientific literature. A lot more has been discovered about the early history of life in the Precambrian. All this goes to show that palaeontology, the study of dead organisms, is, paradoxically, very much alive.

I have also taken this opportunity to correct several small errors that crept into the first edition, and I thank colleagues who pointed them out to me. I particularly thank Angela Milner for her advice on *Baryonyx*, Chris Stringer for an update on human history, and Bill Schopf for discussion concerning, and new photographs of Precambrian fossils.

CHAPTER ONE

BURIED IN THE ROCKS

Everybody has seen fossils at one time or another, dinosaurs in the Museum displays, ammonites collected from a seaside holiday, or odd-shaped stones picked up on a country walk. The science of fossils is palaeontology, which is a long word for the business of studying the life of the past. This may create an image of a palaeontologist as a fusty old professor, as desiccated as some of the fossils he studies, busying himself in drawers thick with dust. But like most other sciences palaeontology has been growing in the last decades, there have been many new and exciting discoveries and new ways of looking at 'old' fossils. The study of fossils is inextricably linked with the study of the rocks that contain them, and the fossils provide some of the crucial evidence for the history of the last thousand million years. When one thinks of all that has happened in the last two *thousand* years, to be a historian of a thousand million seems an impossible task. Palaeontologists know that there will always be new fossils to discover. This book will try to show some of the things that can be done with fossils, as well as describing the fossils themselves. It will show how the dead objects can be brought back to life, how they are classified and how they can be used to read the details of past climates and the distribution of continents, very different from their arrangement today. So however mundane a small fossil held in the palm of your hand may seem, it can be the key to unlocking some of the most profound secrets of the earth.

Fossils are the remains of prehistoric animals or plants. Usually they are some hard part of the organism, resistant to decay, that has been preserved enclosed in sediment – past life that has been buried with the rocks and entombed inside them. Fossils can equally be the record of activity of animals – fossil footprints, or the tubes and trails of soft-bodied worms that otherwise leave no trace.

The province of the palaeontologist overlaps with that of the archaeologist, but generally the palaeontologist concerns himself with the organic remains of greater antiquity than the recorded history of mankind. A cow that died 500 years ago and is then exhumed as a skeleton is in a sense a fossil, but its relevance is towards an understanding of the agricultural practices of the time. Archaeology is concerned with the history of organized men over the last 5000 years or so. Most of the fossils we shall be discussing in this book are many times older than this, but of course in the sites where the remains of early man are discovered it is quite possible to find the palaeontologist and the archaeologist working side by side.

Fossils are often small and insignificant, but they can be spectacular, and it is likely that they attracted the curiosity of men from the earliest times. They have been found as 'charms' in the caves occupied by our remote ancestors. It is a remarkable fact that their significance has been appreciated for less than two centuries. Compared with physics or mathematics, palaeontology was a science that had a very late development.

Ancient Greek philosophers made observations on the formation of rocks that seem with hindsight to have been remarkably perceptive, but these early investigations do not seem to have excited the curiosity of the Renaissance scientists. Always an acute observer, Leonardo da Vinci noted the presence of fossil shells in rocks now far removed from the sea and speculated that the rocks must have formed beneath water. His opinions lay largely unheeded, buried in his notebooks. Perhaps the truth is that science had more pressing problems to solve: the nature of motion; the structure of the solar system; the reality of the elements. Until the eighteenth century, observations on fossils were principally the province of the antiquarian who would illustrate such 'figured stones' along with other curiosities of the natural world. Nonetheless collections were being made and housed in universities like Oxford and Cambridge, the raw material for subsequent interpretations.

The attention focused upon fossils was part of the great surge in interest in the ordering of the *living* organisms of the world. Early accounts of living animals and plants had often been fanciful and inaccurate, and there had been little attempt at a rational ordering based upon their shared similarities. During the last part of the seventeenth and the eighteenth centuries all this was to change: it was an age that felt conviction in the presence of order in the world. At the same time the standard of illustration and description of plants and animals improved enormously, partly because there was a market for well-produced books on nature in the libraries of the rich. The work of John Ray (1627–1705) in England, Georges Buffon (1707–88) in France and Carl von Linné (or Linnaeus) in Sweden (1707–78) set out the outline of classification of organisms still in use today. Of pre-eminent importance was the recognition of *species*, the different kinds of animals and plants, which formed the units of classification. Linnaeus evolved the hierarchical classification of organisms that we still employ, and to a certain extent naming something is a step towards understanding it; it gives a common vocabulary, so that men of science the world over know that they are talking about the same plant or animal when they use its scientific name.

It was a natural extension to apply the Linnean system for naming organisms to their fossil equivalents; the framework for their logical arrangement had been established. Linnaeus himself paid little attention to the sort of invertebrate animals that constitute the bulk of common fossils, but Jean Baptiste Lamarck (1744–1829) made a special study of such humble organisms and was able to apply his researches on living animals to the naming of fossil forms. Up to this time it was a relatively open question whether species were 'fixed' or not and

Lamarck firmly believed that one species could change to another by the inheritance of advantageous characteristics. But belief in the fixity of species became more generally accepted, for it was espoused among others by the great comparative anatomist Baron Cuvier (1769–1832), and seemed to be consistent with a religious view requiring the special creating hand of God in organizing the wonderful diversity of the natural world, past and present alike. 'God made them high and lowly, he ordered their estate . . .' as the hymn expresses it.

The science of geology developed as a respectable discipline at about the same time, a product of a similar urge to observe directly from nature and synthesize the observations into principles. Theories about the significance of fossils could not be divorced from ideas about the formation of the rocks that contained them. As so often happens, the early ideas had a compelling simplicity: in the eighteenth century the influential mineralogist Abraham Werner thought all rocks were accounted for by a great event of precipitation from a universal ocean. The coarse-grained rocks (which we now recognize as igneous in origin) were deposited first, followed by sedimentation of all the overlying rocks, resulting in the layering of sedimentary formations. Such an event required little time and was consistent with a biblical notion of Creation. The idea of vast stretches of 'geological time', which now seems familiar enough for even small children to talk glibly of millions of years, would have seemed quite alien then. The origins of our ideas go back to one of the great classics of geology, *The Theory of the Earth*, published by a Scot, James Hutton, in the closing years of the eighteenth century. Hutton based his book on detailed field observations of rocks, and recognized the importance of looking at the formation of present-day sediments in the interpretation of their ancient equivalents. There was nothing particularly mysterious or catastrophic about the past recorded in the rocks; to understand the past it was only necessary to look at the processes at work in the world today. But of course the formation of all these rocks required huge expanses of time, and creation must have been a protracted process. Hutton's ideas took root, but only slowly. In the earlier part of the nineteenth century the most widely accepted beliefs on the formation of rocks were a more sophisticated development from Werner's: the varied faunas and formations

William Smith, the 'father of English geology'

were the product of successive catastrophies, after which the world had been repopulated by new animals and plants, created afresh, to set the stage for the creation of the present world as explained in biblical revelation. The last catastrophe, and not even the most dramatic, was the Flood. Evidence of the Flood could be found in the cave deposits, which were the youngest known (what we would now call Pleistocene), where the remains of bears, rhinos and elephants could be found in areas where they do not live today. William Buckland's *Reliquiae Diluvianae* (1824) documented many of these in impressive detail.

The first half of the nineteenth century was also a time of great practical enterprise, when pure science and technology continually interacted to throw up new inventions of immediate relevance to a burgeoning industrial society. The discoveries of pragmatic men without formal scientific training could proceed independently of the current scientific controversies on theories of the earth. When William Smith published his geological map of England and Wales in 1815 its practical importance was obvious. By mapping out the formations of rocks you could tell the best route through which to push a canal, or where to obtain clay for brick-making. Smith had recognized the importance of fossils in characterizing the

various strata he coloured on his map. He had collected these fossils from the trenches he cut for his canals, and from small quarries and pits cut for lime or bricks, that were scattered over the English countryside. The fossils varied in kind according to the position from which they were recovered; older rocks evidently lay progressively beneath the strata that covered them. And so piece by piece an atlas of fossils could be compiled which identified the strata and the order in which they occurred. Once the pattern was known a traverse into strange ground could be interpreted simply by examining the fossils culled from the local exposures. The zoological relationships of the fossils did not matter too much; in the same way that you do not have to know the name of a criminal in order to be able to identify him from his fingerprints. They were a practical and effective method of solving the problems of the structure and order of the rocks in England. When he was presented with the Wollaston Medal of the Geological Society in 1831 Smith was dubbed 'The Father of English Geology' by the President, a tag that has remained. Similar elucidation of rock succession and the fossils enclosed within them was being pursued in France, and by the 1820s it was clear that fossils were not merely objects to divert the antiquary, but were of great practical importance in solving geological problems, and many of these in turn were of direct economic relevance. Fossils have continued in this role ever since, and in a sense the many palaeontologists employed by oil companies today to identify fossils owe their jobs to William Smith and his contemporaries.

In twenty years from the publication of Smith's map the government too had begun to realize the implications of the kind of geological mapping and collecting that Smith had pioneered, and the Geological Survey of Great Britain was accordingly founded. The early officers of the Survey include some of the most distinguished names in geology. Their successors are far more numerous, for more knowledge always results in new techniques for answering yet more questions. And the more searching that was done, the more new fossils came to light, many of them stranger and more exciting than could have been dreamed of fifty years before. In 1825 the description of the first dinosaur bones was published. So much was discovered that scholars became specialists in the description and identification of particular groups of fossil animals and plants. Money was available to publish great compendia of such descriptions, and most of the fossils were described therein for the first time. In the great age of the amateur naturalist there was a good deal of public interest in such monographs, and local natural history societies seemed to spring up wherever there were enough people to form a committee. The study of rocks became respectable employment for gentlemen of private means. The 1850s to the 1890s were in many ways the heyday of palaeontology, for books about fossils were only exceeded by the discovery of new kinds of extinct organisms. Many of the classics of the subject were written then, and the fossil faunas of Europe and North America were gradually exposed to public view in museums. In the process it slowly became commonplace to think of larger and larger stretches of geological time during which the strange former inhabitants of the earth had their prime, and perished, to be replaced by others.

Geological mapping proceeded over a wider and wider area, and time and again fossils proved their worth in determining the order of succession and identifying the subsequent events that twisted and distorted the rocks into the structures they have today. Such studies changed the way the landscape itself was observed: this high range of hills was the product of a convulsion of the earth which folded the rocks into complicated piles, that valley was the floor of an ancient lake, the water from which had long since drained into the sea. As knowledge increased, so the explanation of the shaping of the earth by a series of catastrophes came to seem excessively arbitrary. James Hutton's ideas of interpreting the past by observation of present-day processes were more persuasive, and received cogent and sweeping support from the works of Charles Lyell, especially his *Principles of Geology* (1830–3). The past became subject to the same natural laws that prevail today, intensified in their operation at some times to be sure, but the mysteries of the rocks could be unravelled by observations from volcanoes, seas, and winds that can be directly recorded.

The validity of this way of looking at things has remained, for all the realization that the most distant past of the earth was different in several important ways from the recent; geologists still look at the formation of present-day sediments to interpret the features of ancient rocks. Those who accepted Lyell's book had also to accept another idea. It was common observation that rocks, which had originally been deposited beneath the sea, now formed huge cliffs, and when the thickness of *all* such sedimentary rocks from the different periods of geological time were piled on top of one another the total must be immense. Yet sediments accumulated slowly, a centimetre or two a year. So it must have taken an inconceivably long time to accumulate this great pile of sedimentary rocks; geological time must be reckoned in millions, not thousands of years. The biblical Creation story could not be literally true. And since fossils of many kinds were found in the sedimentary rocks, life must have been present on earth for a comparable period of time, time enough for the changes between one kind of animal and another to have happened. It was not necessary to have catastrophic extinctions followed by re-creation of new life forms. The living biological world could have originated by a process of transformation, just as the world itself was slowly shaped by the same forces that operated in it today. The marriage of geological time with transformations of one species of animal or plant into another led to one of the simplest, but one of the most important and influential ideas: the theory of evolution.

Charles Darwin published his great book *On the Origin of Species by means of Natural Selection* in 1859, almost thirty years after Lyell's *Principles of Geology*. The title would scarcely recommend itself today as a best seller, but that is what it was, sold out as soon as published. The *idea* of evolution was not in itself a new one; it can be found implicitly or explicitly in many earlier works, notably by Lamarck. The time was propitious for a summary of the kind that Darwin provided, and he described a mechanism which drove the process of change. Phrases that originated (or were garbled from) Darwin's book have entered the language – 'the survival of the fittest', 'Nature red in tooth and claw' (coined by Darwin's champion T. H. Huxley). Perhaps it was this somewhat stark view of nature, not Mother Nature at all, that was partly responsible for the furore that was caused by Darwin's book. There was also the clear implication that man, as another animal, was subject to the same inexorable process. If Darwin was correct the boundary between man, with a soul, and the ape without one

MAN · IS · BVT · A · WORM ·

Darwin's death recorded by *Punch,* showing a naive view of the progress of evolution!

was a slim one indeed. The orthodox clergy became ranged on one side, and many, but by no means all, of the leading scientists on the other, and it was a number of years before the concept of evolution became generally acceptable. There are still religious fundamentalists who regard the battle as continuing. In fact, Darwin shared the first proposition of his theory with another biologist, Alfred Russel Wallace, who had independently reached similar conclusions, although Darwin's book indicated the broader scope of his ruminations on the subject. But this does show that the time was ripe for such a change of outlook on the origins of the diversity of life. That it was Darwin who expressed the new view was certainly no coincidence. By the time that *Origin* was published, he had already written

several papers that were (and have remained) classics of their kind, notably on the origin of coral reefs. He was a consummate observer. Many of his ideas probably had their origin in the world voyage he made on the *Beagle,* and the genesis of many of them are to be found in his diary of the journey, and in his letters. Even after the publication of his most famous book, which would have been enough for most biologists, he continued to make fundamental contributions to our way of looking at the natural world, all of them founded on meticulous observations, in some cases extending over decades. He was the quintessential biologist.

The combination of Darwin's ideas with the expanded geological time-scale gave a new impetus to palaeontology. The rocks recorded the very history of evolution.

Darwin himself recognized the imperfection of the fossil record, and devoted some space in the *Origin of Species* to explain why this was so, but the theory provided a new framework in which to interpret fossil remains. Argument still continues over just how much the fossil record can reveal, but the fact is that diligent search of the rocks has uncovered animals that showed how one group links up with another, and how the post-Darwinian world was derived. Inevitably the search extended for the ancestors and relatives of man himself, and eventually candidates for this role were discovered in Africa, although it took a century to find them. Advances in other fields of biology continued, and they in turn have been assimilated into the palaeontological perspective, but the mode of operation of the science was well-established by the close of the nineteenth century. Detailed investigations into the process of evolution as reflected in the rocks are continuing, and many of the questions on the evolutionary origins of living animal and plant groups are as hotly debated today as they ever were. But few people doubt the overriding principle of evolution in the shaping of the biological world.

TURNED TO STONE

When fossils were still regarded as curious freaks of nature, explanations of the way they were preserved could be fanciful. Ammonites found at Robin Hood's Bay on the Yorkshire coast were supposed to have been snakes turned to stone by St. Hilda. Enterprising local craftsmen embellished the fossils by carving on an appropriate snake's head! (*see* above). As it became clear that the fossils were there in the rocks as the result of natural processes, the mechanism of fossilization became better understood. The great majority of fossils are of hard parts of the animal or plant, or structures resistant to decay. Shells, bones, wood, and teeth are all likely to be preserved as fossils. Molluscs and mammals are thus likely to have a good fossil record, worms and amoebae a poor one. The clusters of dead shells one finds in rock pools on the beach – limpets, crab claws, a winkle or two, a broken sea urchin – are typical of the sort of debris that may become fossilized. Many fossils result from the cast off, outgrown or broken shells of marine animals. All that is now required is for the shell material to get covered with sediment. In a sense it becomes a fossil from that stage on, and many quite ancient fossils a million

years old or so can look remarkably like shells picked up from the beach today. Most shells are more or less porous, and bone is very spongy, and so very frequently the small pores in the fossilized material become the site for deposition of minerals, which make the fossil more dense than the shell of the recent animals.

As sediments pile up at the site of deposition their coherence increases, some of the water they contain is expelled, and many more subtle chemical changes go towards making the rock types with which we are familiar: shale, sandstone, limestone and the like. In the harder rocks, such as limestone, the fossils buried with them retain much of their original shape and convexity, but shales frequently become compressed, and as this happens the fossils become flattened (*see* p. 15). Sometimes the same fossil species can present very different appearances according to whether it has been preserved in the round, or flattened. Further changes can happen to the fossil as a result of their sojourn in the rocks. Most shells are made of calcium carbonate, which is soluble in carbonated water. In porous rocks, especially sandstones through which ground water courses, the shell itself can be dissolved away. The fossil is not lost in this case

An ammonite (*Dactylioceras*) from the Yorkshire Jurassic rocks, with a snake's head added by a sculptor to show the miraculous hand of St. Hilda

A flattened Cretaceous ammonite (*Schloenbachia*). As the sediment is compressed the fossil loses its original shape. The flattened example is on the right.

Internal (left) and external moulds of a trilobite

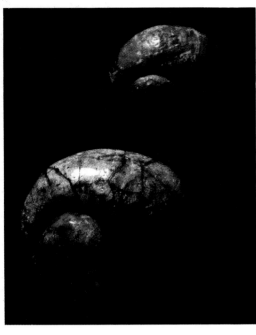

because the rock itself will have taken an impression of the shell, just as a fingerprint can be impressed into clay. When such a rock is split open (*see* p. 15) the inside of the shell will be preserved as an *internal mould*, while the 'other half' (the *counterpart*) of the specimen will contain the impression of the exterior details. Obviously, to obtain a reconstruction of the whole shell you need *both* halves of the fossil specimen. The golden rule of collecting is: *Never throw away the counterpart!* Many a vital specimen has gone tumbling down a scree slope when this rule has been ignored. In other cases the cavity left after the shell has been dissolved is then replaced by another mineral; if you are very lucky the mineral might be opal (*see* above).

Sometimes these secondary replacements can be turned to the advantage of the palaeontologist. Fossils with shells originally of calcium carbonate may be replaced by the mineral silica. If this happens in a limestone, the enclosing rock can then be dissolved in acid, and because the silica is insoluble, the replaced fossils dissolved out of the rock. In this way remarkably delicate details of the original fossils can be preserved: spines and other features, which are impossible to dig out mechanically, are easily observed on such specimens (*see* p. 17). A rarer type of preservation is perfect replication almost molecule by molecule of the original fossil, so that minute details of the microstructure are determinable. Such petrifactions are usually in fine-grained silica. One famous example is the Rhynie Chert, a Devonian petrifaction of some of the earliest land plants. The perfection of preservation is such that sections of early plants clearly show the individual cells in the plant tissues (*see* p. 17). Such specimens are of immense importance in revealing the most intimate details of the structure of extinct organisms, which can then be compared with some confidence with their living relatives for which there is complete information.

GEOLOGICAL MIRACLES

In very rare cases the rule that the fossil record preserved only the hard parts of animals no longer applies. These geological 'miracles' also preserve the impressions of the soft organs of animals that would normally decay without trace. There are always special geological circumstances in such cases. One of the most famous, and oldest, of these is the Burgess Shale from British Columbia, Canada. This is a black, fine-grained Cambrian shale with a host of wonderfully preserved fossils, many of which are unknown anywhere else. It affords us a unique glimpse of almost the whole spectrum of life at this very early stage in its history, and shows how many animals are *not* preserved under the normal circumstances of the fossil record. Trilobites and other arthropods are here preserved with all their limbs, antennae, and even their gut contents (*see* p. 18). Some of the animals are hard to match with *any* living organisms, and may represent kinds of creatures that have long since vanished from the earth, early experiments that were unsuccessful and left no other trace. Many kinds of 'worms' are present in the fauna. The soft parts are preserved as the thinnest of films, and the whole fauna was probably buried rapidly, before the soft parts could decay, and in an environment where they could not be shredded to pieces by scavenging organisms. A similar occurrence in the Devonian Hunsrück Shale has the soft parts preserved as a thin film of the mineral iron pyrites. In this case the structure of the soft parts can be studied by means of X-rays which pick out the iron pyrites – the structures can be photographed in the rock even when they are not visible on the surface of the rock! (*See* p. 19.)

The early bird *Archaeopteryx* is preserved in a fine-grained limestone, creamy brown in colour, which was used for the manufacture of lithographic blocks, found near the Bavarian town of Solenhofen. This could scarcely be more different in appearance from the Burgess Shale, but like it the

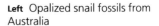

Left Opalized snail fossils from Australia

Right Section through a plant stem from the Devonian Rhynie Chert, Scotland, showing how even individual cells may be preserved

Silicified brachiopods, showing long, delicate spines which would normally break off in the rock

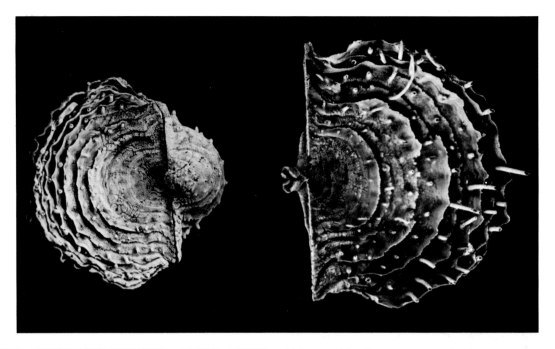

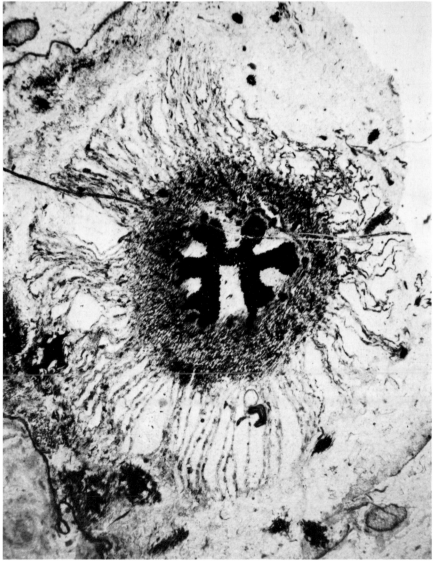

lithographic limestone retains spectacular remains of a whole host of animals with scarcely any fossil record elsewhere. Being Jurassic in age, many of these fossils are related, distantly, to animals still living. Besides the famous earliest bird there are reptiles, dragonflies, crabs, relatives of the King Crab *Limulus*, sea spiders, and mammals. This is a mixture of terrestrial and marine life. The Solenhofen deposits are supposed to have accumulated on the boundary between land and sea, probably in a lagoon, where a sticky, limy mud was accumulating. Flats of this mud were probably exposed at low tide, and at this time *Archaeopteryx* became entrapped. The fine mud was ideally suited both to entomb the remains and to take an impression from delicate feathers that would otherwise have decayed without trace. It is fortunate that this happened, because otherwise there would be doubt whether or not *Archaeopteryx* was really a bird. Some such consideration has even led to the absurd claim that the *Archaeopteryx* specimens were fakes, manufactured by impressing feathers of living birds on a prepared surface surrounding genuine small dinosaur fossils. Quite apart from the reflection on the probity of the curators who have studied the fossils, this claim is inherently improbable given the discovery of fossils of the first bird from time to time over many decades (it would imply a scientific conspiracy of massive proportions, and without commensurate motivation); nor has careful examination of the fossils revealed any evidence of

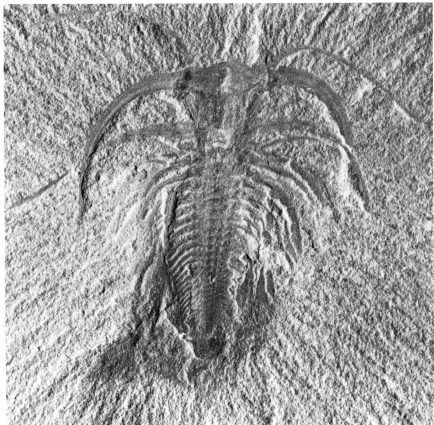

such fakery. Geological miracles like the Burgess Shale and the Solenhofen limestones are of a palaeontological importance inversely proportional to their rarity – the information they yield is like a floodlight on the past, when most geological sites are more like an intermittent flashlight. The geologically youngest of such exceptional fossils are the frozen mammoths of Siberia, dating from late in the last Ice Age. The extinct giants were apparently frozen so quickly that their hair and meat are preserved almost as it they had been frozen for the supermarket. Speculation about regenerating one of these remarkable animals from their frozen cells are (however) probably over-optimistic.

Amber ornaments were extremely popular in Victorian times, and the most prized of these had small insects displayed within the amber droplets. Apart from its beauty, amber preservation is another exceptional occurrence where animals not normally preserved as fossils are found in abundance. Amber is generally 70 million years old or less (some is as old as 100 million years), and the majority of the insects are related to living forms, but they are of great importance in understanding the genesis of the most diverse group of living invertebrates.

Amber started as a resin oozing from the branches and trunks of coniferous trees. Insects and spiders were trapped in the resin, and enclosed within it as more resin was added to the droplets. Hardened resin is extremely tough, and so has a high chance of being fossilized, carrying with it its cargo of preserved insects. Lumps of what is now amber eventually found their way into sediments, from which they can be recovered like any other fossil. The same process of entrapment still goes on today, and sometimes the Recent hardened resin (copal) can be sold as ersatz amber; it is usually much lighter in colour, however.

TRACE FOSSILS

A fascinating branch of palaeontology is concerned with the traces left by the activity of extinct animals – trace fossils (*ichnofossils*). These can be footprints, like the tracks of dinosaurs (*see* p. 20), or the trails left by animals as they grazed the sediment looking for food, or sometimes the burrows made while the culprit was escaping from a predator or laying its eggs. In many cases the animal itself is not preserved, and although it may be possible to deduce what it was doing, it is not possible to say what kind of animal it was that was responsible for mak-

Above left A trilobite (*Olenoides*) from the Burgess Shale, Cambrian, western Canada, showing the traces of its limbs and other soft parts not normally preserved in the fossil state.
Above *Marrella*, also from the Burgess Shale, a peculiar primitive arthropod unknown elsewhere. *Photographs kindly supplied by Professor H. B. Whittington*

X-ray of a magnificently preserved starfish from the Devonian Hunsrück Shale, Germany. *Photograph by Professor W. Stürmer*

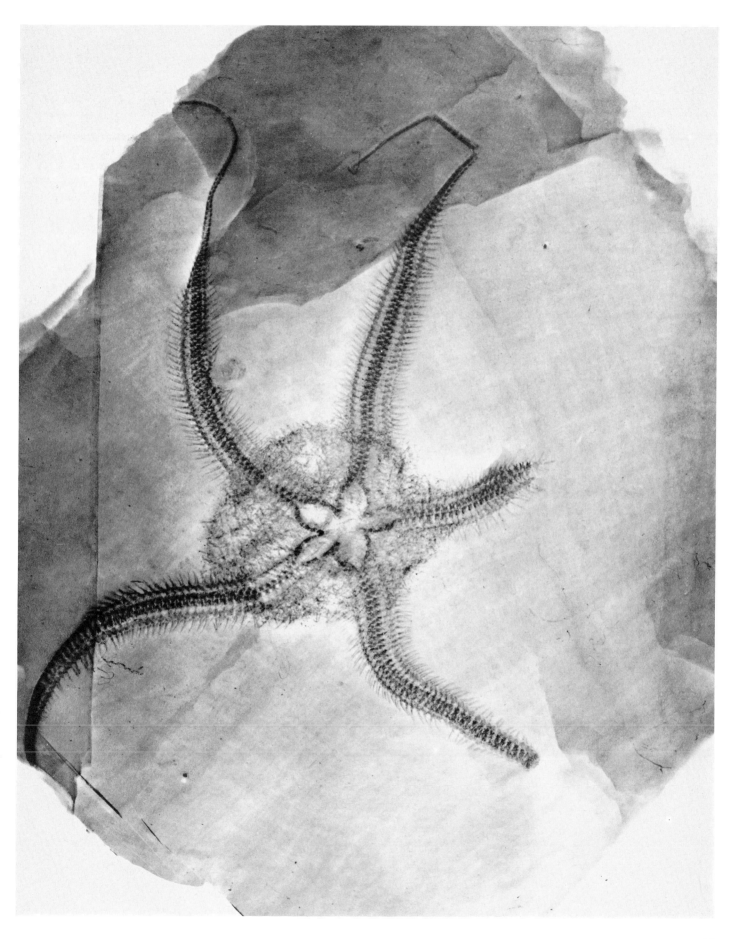

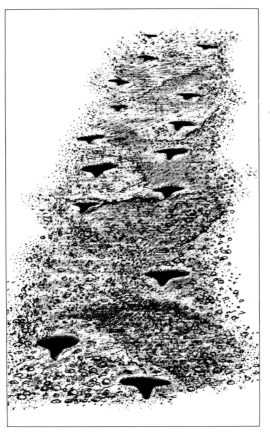

Right The three-toed tracks left by a dinosaur

ing the track. Huge, three-toed tracks of dinosaurs (*see* left) can tell us a lot about the gait of the animal that made them – what its stride was, whether or not the front limbs touched the ground. Many tracks and trails are rather inconspicuous, labyrinthine or braided paths made by worms. Some of the worm 'castings' are about the only fossil record we have of these creatures. Trace fossils are in abundance even in the early Cambrian. They are especially numerous in sandy rocks, which otherwise lack body fossils. One of the Cambrian occurrences is the famous 'Pipe Rock' of the northwest Highland: the pipes that give the rock its name are closely packed straight tubes which were made by some sort of 'worm'. Other beds of rock in the same formation contain U-shaped tubes or funnel-like ones. The tracks made by trilobites are sometimes numerous in rocks of Cambrian and Ordovician age. These include winding trails, and short burrows looking like a pair of beans (*see* below), on some of which the impressions left by the legs of the trilobite are clearly visible. Trace fossils are given names, just like body fossils; the trilobite tracks are called *Rusophycus* or *Cruziana* for example.

Below *Rusophycus* – the burrows dug out by trilobites.

'Worm' tracks, filled in with sediment of a slightly different colour from the surrounding rock

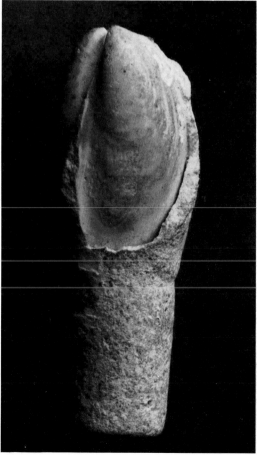

A fossil boring bivalve mollusc, preserved at the end of its boring, about twice natural size

A little explanation is needed about the preservation of tracks: most tracks are dug into the sediment surface; an overlying layer of sediment then fills up the tracks, so that the cast formed by this infilling is a *positive* impression of the track itself. Other burrows are dug deeper into the sediment, and they often fill, once vacated, with more sediment of a different colour. Collecting past tracks can be almost as instructive as collecting the fossils of shells. Different kinds of animals live in different environments, and leave differing evidence of their activities; so by studying tracks it is possible to find out a lot about the animals that lived on the former sea floor even without remains of the animals themselves. Tracks should be distinguished from *borings*. These are made into a hard medium – such as rocky surfaces or wood. A number of organisms are adapted to what might be termed (without impoliteness) a boring existence. Some bivalved molluscs specialize in cutting into hard rocky surfaces, and these, too, can be found as fossils, usually sitting at the end of their self-made caves (*see* left).

HISTORY OF FOSSILS WITHIN THE ROCK

Once they are incarcerated in the rocks

Angelina, a trilobite, which has been distorted, by stretching of the rock that contains it

fossils are passive passengers, and what subsequently happens to the rocks also affects the fossils themselves. Not all rocks lie undisturbed until the hammer cracks open the booty they contain. Initially the fossils usually lie parallel to the *bedding planes* – the more or less horizontal surface representing the former sea floor. Many fossils remain in this attitude as the rocks are eventually uplifted above the former ocean bed, to become exposed in cliffs, quarries or cuttings. More usually the uplift process, which may be connected with earth movements (*see* Chapter 6), also produces tipping and gentle folding of the rocks, so that the bedding planes are no longer horizontal (*see* p. 23). This does not affect the fossils, which can be collected in the usual way. It is not uncommon to find the rocks tipped vertically. But where the earth movements are more violent, the rocks may be squeezed and distorted, and so are the fossils contained inside. This is particularly the case in soft rocks, like shales. Older fossils are more likely to be found in this condition, because they have had more time to be involved in violent events. But many of the most ancient fossils have escaped such accidents, and are as well-preserved now as they were when they were first fossilized. Distortion includes stretch-

ing (*see* above) and twisting. At the same time a lot of the finer details are destroyed. Obviously palaeontologists would like to have perfectly preserved material to work from, but often distorted fragments are all that is available from a large area, and they have to be identified as best they can.

The process of distortion does not stop with mere stretching and bending. As the rocks are squeezed progressively they sometimes become heated strongly and under these conditions the rock itself begins to change. This can have the effect of removing the fossils completely, the small ones first, then the most robust. Even so, fossils have been known to survive the most intense heating and high pressure. A very common secondary change that happens to shaly rocks under pressure is that they develop a *cleavage*. This means that they start to split not along the bedding planes on which the fossils lie, but at a high angle to the bedding, sometimes at right-angles. The kind of black or purple slates used for roofing are cleaved rocks of this kind. You can often see stripes passing across such slates: these are sections through the original bedding. Obviously it is no use splitting open such rocks to find fossils parallel to the cleavage – you might see a section through one if you are lucky. Fossils can be recovered with much hard work by smashing the slates in such a way that part of the original bedding is exposed. They are usually in rather a sorry state by the time they are found.

BITS AND PIECES

Collecting a pile of fossils is only the beginning. Many fossils are only fragments of the whole animal or plant. To piece together the whole organism is rather like doing a jigsaw puzzle without the benefit of the complete picture to work towards. Piece has to be added to piece, and the larger and more fragmentary the animal the more the result is in question. Not surprisingly mistakes have been made. The first reconstruction of the dinosaur *Iguanodon* finished up with its thumb on its nose! Trilobites are much more common as pieces than as whole animals, and many kinds are known only from heads or tails until one day a lucky collector turns up a whole one. The problem is particularly acute for large plants, because there is nothing very obvious to connect the root with the trunk, or the trunk with the leaves if they are preserved in different places, and

they usually are. Flowering structures, not being easily fossilizable, are sometimes even more difficult to assign to the whole plant. The result is that different names are given to different organs, one for the root, one for the bark and so on. Eventually, when the links are made, one name suffices for the whole plant, but it is sometimes handy to be able to compare leaves, say, by using the different 'leafy' names.

There are many problems with piecing together fossils: distorted fossils have to be restored to their original shape, left valves of clams have to be matched with right, vertebrae and limb bones placed together in the correct order. Many fossils are only known from fragmentary material, and it may be years (if ever) before the next piece is discovered. Just as there are rare and abundant species today so there are rarer and commoner species of fossils. Some species are so abundant that one is almost certain to turn up a specimen if rocks of the right age are hammered, others are so rare that a collector may go to a quarry one day and find a specimen, and visit the same place for years afterwards without finding a second one. This is all part of the particular fascination

Folded rocks like these may contain fossils that have been variously distorted from their true shape

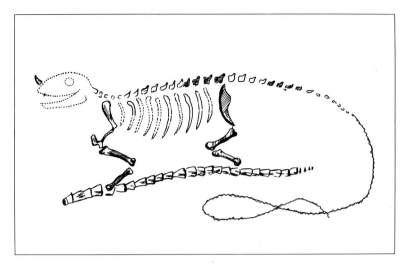

Gideon Mantell's original restoration of *Iguanodon*, compared with a modern version of the same animal, model (*below*)

of palaeontology; you never know quite what will turn up. The discovery of new kinds of fossils is a regular occurrence, even in ground that has been traversed by many collectors before. And each new discovery patches in a little of the evolutionary story that was previously obscure. Even now many parts of the world are being explored for the first time, and the new palaeontological finds show little sign of abating. Professional palaeontologists often feel swamped by the sheer variety of different fossils there are to deal with. It is only a question of searching.

Although they are fascinating just as attractive objects, fossils are of enormous practical importance in interpreting the history of the planet as a whole. As we have seen in this chapter, the study of fossils links in with other branches of geology and biology, indeed a knowledge of related sciences is essential to appreciate the significance of fossil finds. The next two chapters will explore some of these connections, to show how the animal and plant life of the earth is intimately bound up with the story of the rocks themselves, and the very configuration of the continents and oceans. As Lyell recognized, the forces of physics have been the same through geological time, so the processes that formed the present earth are not beyond our understanding. But the unchanging physical laws operate on a thoroughly mutable world, and the configuration of land and sea has changed repeatedly. For hundreds of millions of years living organisms have altered in harmony with the world, and in the process have themselves transformed it.

SETTING THE STAGE: TIME AND CHANGE

The earth is 4500 million years old. This figure is now generally accepted, but has only become so in the last few decades, especially since the age of the moon was determined. If we return to the nineteenth century, at the time when Hutton saw that the earth must be 'immeasurably' old for present processes to account for all the varied features of the rocks, and their immense thickness, the business of putting an actual age on the earth was very problematic. Once the idea of Creation within a few thousand years disappeared, it was obvious that the earth had to be much older. But how much? Millions of years certainly, but ten million, a hundred, or a thousand million? The fact is that it is impossible for the human mind to grasp such lengths of time as these. It was easy enough to realize that a long span of time was needed, but difficult to devise methods of assessing the time involved.

Once the *sequence* of rocks was worked out, it was possible to add all the piles together to say something about what had happened through geological time, the different kinds of organisms that had populated the earth and replaced one another. But this still left the underlying Precambrian rocks, which seemed to be barren of all fossil life. How much of the earth's history did these represent? At least one school of thought believed that these ancient rocks were partly the original crust of the earth (many of them being crystalline) – in which case an approximation to the age of the earth could be obtained by guessing at the length of time of deposition for the Cambrian and later rocks. If rocks accumulated at a fixed rate, then the maximum total thickness of rocks should be a basis to measure the time

it had taken to deposit them. An estimate of 500,000 feet of post-Cambrian rock is not unreasonable, and since a fast rate of sedimentation is about one foot over 500 years today, we could get a figure of about 250 million years for the whole pile – not a bad estimate. However at the time the subject was debated in various ways: how much thinner had the rocks become by burial, and similar questions were raised, and the estimates ranged from less than 20 to more than 700 million years. The great physicist, Lord Kelvin, had found another method of making the estimate. He assumed that the earth had cooled from a totally liquid state, and if this were the case then the time taken to produce the condition of heat flow and solid crust we have today should be between 20 and 40 million years. By the 1890s this seemed to be the most 'scientific' answer.

At the same time the knowledge of the *relative* time scale of geology had come to resemble closely the scale we use today. The question of the *absolute* (in millions of years before the present) ages was a different one, and the whole edifice of geology was built largely on relative relationships. Obviously the Silurian rocks were older than the Devonian ones, because they underlay them, and the fossil fauna could be shown to be different; the Devonian in turn underlay the coal-bearing Carboniferous, and these rocks again the Permian formations. Such a relative scale was implicit in William Smith's map, and, accompanied by wrangling and argument, the whole succession of fossil-bearing rocks was elucidated step by relative step.

The history of the earth was written in the rocks, and the rocks apparently were divided into natural units. The fossil faunas

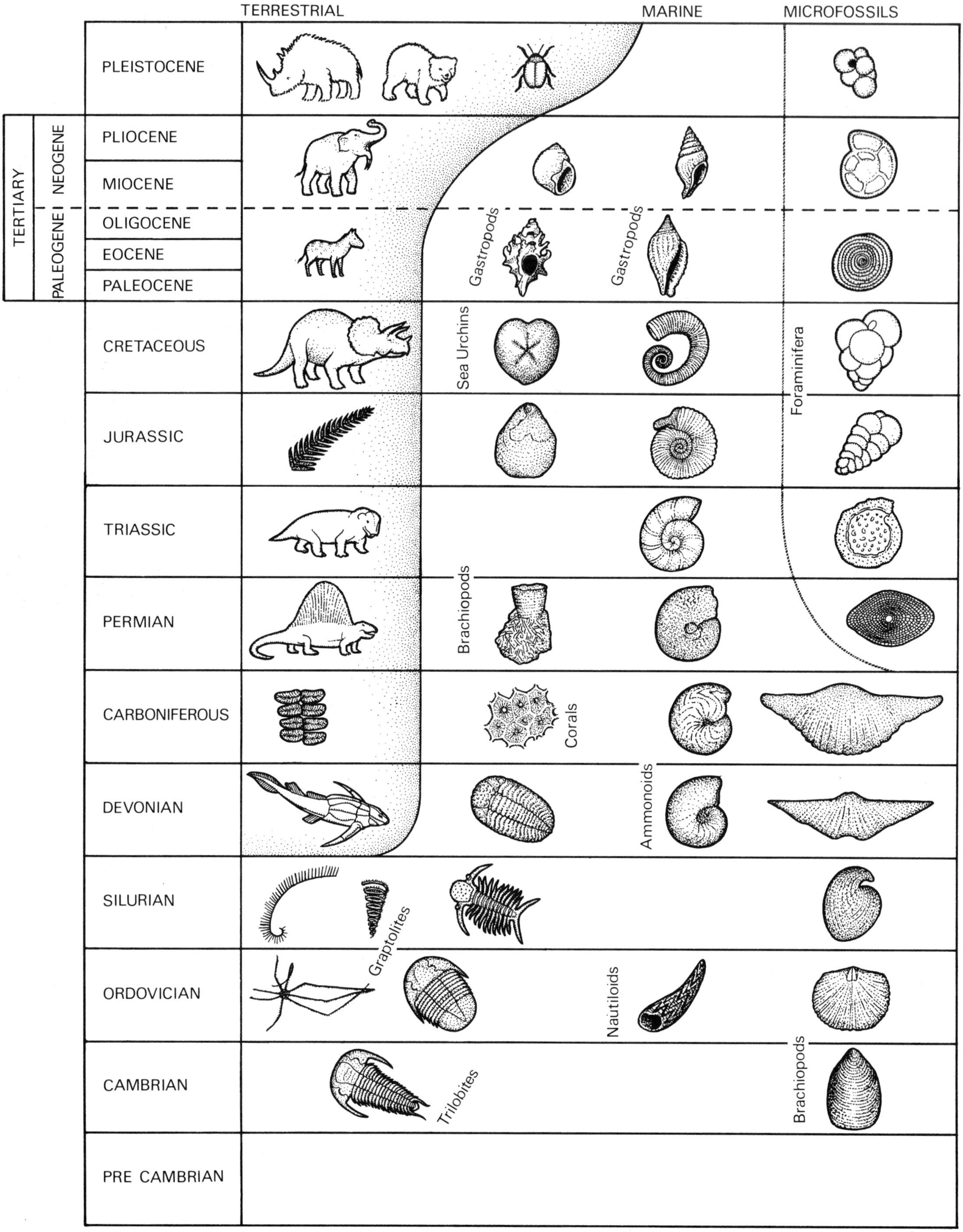
TERRESTRIAL
MARINE
MICROFOSSILS
TERTIARY
NEOGENE
PALEOGENE
PLEISTOCENE
PLIOCENE
MIOCENE
OLIGOCENE
EOCENE
PALEOCENE
CRETACEOUS
JURASSIC
TRIASSIC
PERMIAN
CARBONIFEROUS
DEVONIAN
SILURIAN
ORDOVICIAN
CAMBRIAN
PRE CAMBRIAN
Gastropods
Gastropods
Sea Urchins
Foraminifera
Brachiopods
Corals
Ammonoids
Graptolites
Nautiloids
Trilobites
Brachiopods

also seemed to change in a radical way from one of these broad units to another, and what better way, therefore, to define the divisions of geological time? The development of the formations of rocks varied obviously from one area to another, and in some cases provided the name of the geological period: the Jurassic period with the Jura mountains, for example. In the first instance some of the natural divisions of geological time were bounded by *unconformities* (*see* above). Rocks lying beneath an unconformity have obviously had time to be uplifted and folded before the rocks lying on top of them were deposited. The divisions of geological time from the Cambrian onwards were established in Europe, and it is a measure of the acumen of the geologists who established them that the original concepts survive in all their essentials today.

The fifteen major divisions (*periods* or *systems*) are shown on p. 26. The first four of these (Cambrian, Ordovician, Silurian, Devonian) all take their names from British localities; Cambrian from the Latin word for Wales, Ordovician and Silurian from two of the old Welsh tribes, and Devonian, of

course, from Devon. The Carboniferous were the coal-bearing strata. Permian rocks were typically developed in the Perm mountains of Russia. Triassic rocks naturally fell into three (Tri-) divisions in the characteristic development in Germany, and the Jurassic period was christened as mentioned above. The Cretaceous included the pure white limestone known as *chalk* (Latin: creta). For broader purposes it is still useful to talk about *eras*, three great divisions of fossil-bearing time: Palaeozoic (literally 'old life' – encompassing the Cambrian to Permian periods), Mesozoic ('middle life' – Triassic to Cretaceous) and Tertiary taking us up to the base of the Pleistocene.

The boundaries between the eras represent the most important extinction events the world has known, when much of the living world was replaced by new and different organisms, a graphic way of dividing the history of life into three portions. We should now add a fourth era, the Proterozoic (ancestral life) for the later part of the Precambrian from which fossil organisms are now widely known. The Tertiary periods are shorter than the preceding ones, and have a slightly different basis, being originally proposed by Lyell on the basis of how closely the fossil faunas resembled those of the present day – Eocene ('dawn of the present'), Miocene ('less than present') and so on. Finally the Pleistocene ('nearly present') includes the ice-formed deposits that have been draped like an irregular blanket over the earlier geology of the northern hemisphere, where the geological record merges into the recorded history of man.

Once the language of geological time had been developed it was a great help in communicating the discoveries that had been made internationally, and in building up the first pictures of whole faunas for particular periods. Certain periods acquired popular descriptions, some of which seem to have stuck. The Devonian became the 'Age of Fishes', the Carboniferous the 'Age of Amphibians' and the Jurassic–Cretaceous the 'Age of Reptiles'. These tags have endured, and they do serve a purpose to emphasize some of the largest animals that lived in the respective periods. But they also serve to give a rather false impression of all the different biological events that were going on at the time, as if nothing else mattered but the large and cumbersome reptiles of the Jurassic and Cretaceous. On the other hand the 'Age of Ammonites' does have a less romantic appeal! There is always a temptation to view the fossil record as if it were a kind of staircase with progressive steps leading upwards to man. This is misleading, because evolutionary activity has been unremitting in even the humblest of creatures, and dominance of the natural world by the largest is only a matter of their conspicuousness.

The geological periods were soon to prove only the most general way of sub-dividing geological time. The periods themselves could be subdivided into segments which would enable a much more refined way of talking about the relative ages of organisms. For example, dinosaurs changed greatly during the Cretaceous Period, and it was necessary to have a way of describing the timing of these changes. Periods were divided into Upper and Lower parts (Middle parts in many cases as well). The finest subdivision became (and remains) that of the *Zone*. A zone is a division of geological time characterized by a particular assemblage of fossil species. Recently palaeontologists have distinguished a number of different kinds of zone, but the broad definition given here applies in most cases. A zone is a small segment of geological time through which the evolutionary history of various organisms pass; a number of the animals or plants are unique to the zone, although others which have evolved more slowly may range through more than one zone. The name of the zone is taken from one of the most characteristic of its defining organisms. So, for example, in the Ordovician, graptolites are of importance in subdividing the rocks, and the Zone of *Nemagraptus gracilis* is named after one species which is of widespread occurrence in its zone, although it is accompanied by other species characteristic of the same time period. The zone is another way of communicating the exact age (on the relative time scale) of a fossil, and, one hopes, means the same to a Chinese as it does to an Englishman. A stack of zones, one after the other, can pare down time to thin slices, giving a close control on the relative ages of fossils. Some kinds of organisms have become more useful in the definition of zones than others, and the most useful have proved to be those which

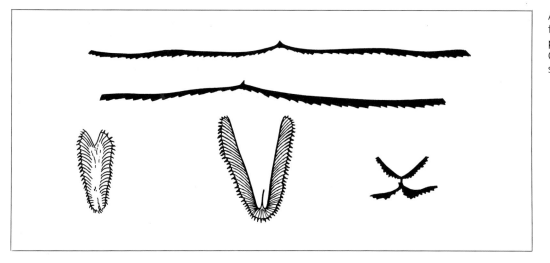

An assemblage of graptolite fossils which can be used to prove a particular zone, in the Ordovician for the assemblage shown here. Natural size

Time measured by radioactive
decay. The half life is when half
the original radioactive element
has decayed away

evolved rapidly (naturally enough, because they enable the time to be sliced finely) and which achieved wide geographic dispersal. It is not much use having a zone assemblage which can be only recognized in one place!

Much of the attention of geologists is devoted to trying to establish the age relationships between strata from widely different localities – the *correlation* of rocks – which is the cornerstone of the branch of geology known as *stratigraphy*. It is not necessary to know the exact age in millions of years to be able to correlate – the basic statement is 'this rock is the same (or not the same) age as that one' whether the age of the rock in question is two or two hundred million years. The fossil content acts as the clock.

The kinds of fossil organisms used as the basis for zones varies through geological time, as one group rises to prominence only to be replaced by another. Strangely enough it is the humblest of fossils that are often the most useful as the basis of the zonal schemes. In theory *any* organism with a good fossil record can be used for a zonal indicator, but obviously dinosaurs are far too large to be recovered from an average roadside exposure even though they may have evolved very fast. Common fossils like ammonites, brachiopods and trilobites are used as zonal fossils, partly because they can be recovered from most (marine) sediments of the right age, and partly because they show enough variation through time to be readily recognized in the laboratory by the palaeontologist. A whole separate branch of palaeontology has grown up around using the smallest of fossils as zonal indicators – *micropalaeontology* – which will be described further in Chapter 8. Small fossils are of particular use in dating rocks recovered from boreholes, where the narrowest of cores may yield large numbers of diagnostic fossils. Not surprisingly, this kind of palaeontology is much employed by oil companies and other commercial enterprises concerned with recovering mineral wealth from considerable depths. Dating the rocks, and correlating between boreholes, is a most fundamental part of mineral and oil exploration. It is perfectly possible for several zonal schemes to exist side-by-side, referring to the same time period; one may be primarily concerned with microfossils, another with brachiopods or ammonites and so on; above all zonal schemes are designed to be of use. One system usually becomes the 'standard' for a particular geological period, often the one first proposed.

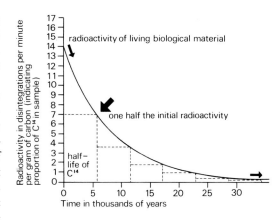

Through the geological column in the table on p. 26 the most important fossils used as the basis of zones change through the column. In the Cambrian, trilobites are more widely used than any other group, largely because they were among the most varied and numerous of the invertebrates at the time. During the Ordovician and Silurian periods graptolites have been widely employed for zonal purposes (*see* p. 28), although trilobites, brachiopods and other fossils are still used in rocks where graptolites are absent. Before the end of the Devonian the ammonoids had evolved, and without doubt they are the single most important zonal group from the Carboniferous through to their extinction at the end of the Cretaceous, although their use is often supplemented by other invertebrate organisms. Micropalaeontological zonations become progressively more important through the later part of the Palaeozoic and the Mesozoic, and the use of the small, unicellular microfossils known as foraminiferans outweighs that of other kinds of fossils in the Tertiary periods. These zones are capable of dividing time into fine slivers. It is estimated that at their best a zone (or its ultimate subdivision, a subzone) can divide time into slices of about half a million years or even less. No other dating method can compare with this for accuracy. Fortunately for the employment of palaeontologists, their services will be required for the estimation of the age of sedimentary rocks for the forseeable future.

RADIOMETRIC AGES

Obtaining an age for rocks in millions of years is now quite a routine geological method, and provides the definitive answer to the question of how long the processes of shaping the earth had taken, that caused so much debate at the end of the nineteenth century. The method depends on the varieties of elements known as *radioactive isotopes*. These are unstable forms of elements that *decay* (change) slowly into other elements, in the process giving off radioactive emanations like gamma rays. The rate at which this happens is controlled by a simple natural law, so by knowing the amount of the parent element left, and the daughter element generated, it is possible to calculate the time taken to produce what we see today. The radioactive clock, unlike fossil ones, converts immediately to millions of years (*see* right). The method depends on accurate measurement of numbers of atoms, and techniques for doing this have improved markedly over the years, so making the result more reliable. The radioactive 'clock' is often set to zero when a liquid rock cools below a certain point: many radiometric ages are obtained from igneous rocks, which of course, do not contain fossils, so the method often operates where fossils cannot be of direct use.

The methods originally employed depended on a few of these natural transformations, particularly that of involving one isotope of uranium changing to another of lead. More and more methods are in use today, and these can now be used on sedimentary as well as igneous rocks. They are often referred to by the names of the mother and daughter elements, such as the important rubidium–strontium method, or the potassium–argon method. The rates at which the decay occurs vary from one natural reaction to another: the slower the process the more use will that reaction be for dating extremely old rocks. On the other hand the change from carbon-14 to nitrogen-14 happens, geologically speaking, quite quickly, and this method is responsible for the so-called radiocarbon ages, applied only to Pleistocene and younger events. At this level, particularly with modern refinements, the method can give a resolution which exceeds that of any other technique, and often is the only recourse for dating isolated sites (like the abandoned camp fires of our forbears). At the other end of the scale the application of radiometric dating techniques has revolutionized our understanding of the huge areas of Precambrian terrain from which fossils are unavailable. And the dating of meteoritic and moon rock has provided the probable age of the earth given in the first sentence of this chapter.

Fossils operate best in the middle of this rather unequal sandwich, and that is the period with which this book is primarily

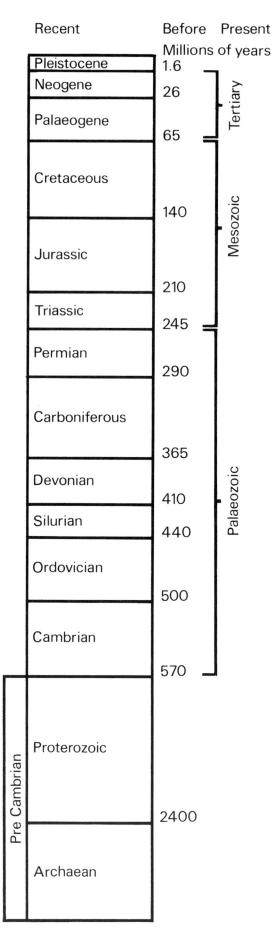

The geological time scale linked into the ages given by radioactive decay

concerned. Here ages derived from radio-active decay of elements provide a scale into which the stratigraphic ages of fossils can be linked (*see* p. 30). Radiometric ages always carry a margin of error ($\pm$) of some magnitude, and certainly more than the length of time for a fossil zone. So the relative time scale provided from fossil evidence is more accurate, and the use of fossils is not likely to be displaced by radiometric ages. The two methods are really complementary: fossils used in the fine division of periods, radio-metric ages as a guide to exactly how long ago events took place, and as a method of dating formations from which no fossils have been obtained. The cooling ages in igneous rocks also give the time at which particular bodies, for example of granite, were *intruded* into the surrounding rock; and where rocks have been reheated and folded during mountain building events this too can 'set' the radioactive clock, providing a method of determining when major phases of metamorphism occurred. All these dates can then be placed back into the geological periods, which provides the general frame-work for talking about the timing of events. The combination of radiometric ages and the relative stratigraphic scale is one of the major achievements of geology in the last decades.

MAJOR CYCLES IN EARTH HISTORY: MOVING CONTINENTS

The evolution of life cannot be separated from the evolution of the planet which is its cradle and its grave. It is now certain that the surface of the earth itself has changed its configuration several times. The earth's crust is a thin skin, of almost negligible thickness compared to the mantle that lies beneath it. The crust moves, and, as it does, geography changes. This happens at an immensely slow rate (a centimetre or two a year) hence the term *continental drift*, but cumulatively the effects are profound. For example, at present Africa and South America are moving further apart. The blocks of continental crust that form these huge areas are thick, tough, and relatively stable. The crust that floors the oceans is relatively thin, mostly composed of basaltic volcanic rock. As the continents drift apart new ocean crust in the form of volcanic rock wells up from the mantle beneath along the mid-oceanic ridges (*see* below). It is appropriate to regard the earth's crust as composed of a number of more or less rigid *plates*, drifting past, towards or away from one another. As these plates are generated along the mid-ocean ridges, so other plates have to be destroyed elsewhere. Again, the thin oceanic crust is what is destroyed, this time along *subduction zones*, where it plunges beneath the relatively rigid continental blocks (*see* below).

An important subduction zone lies off Japan today. As might be anticipated, this area is one of sudden earthquakes, for the downward slide of oceanic crust is far from smooth. Rocks take a great deal of strain, and then 'give' suddenly to the accompaniment of falling buildings and severed pipe-lines. The physical expression of these downward-diving plates is found in the deep ocean trenches off-shore from the continental blocks against which the plates are foundering: great gashes in the ocean floor. Along the line of origin of new oceanic

Generation of new oceanic crust occurs at the mid-ocean ridges. Destruction of oceanic crust at a plate margin down a subduction zone

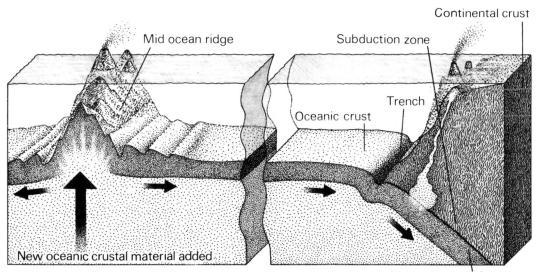

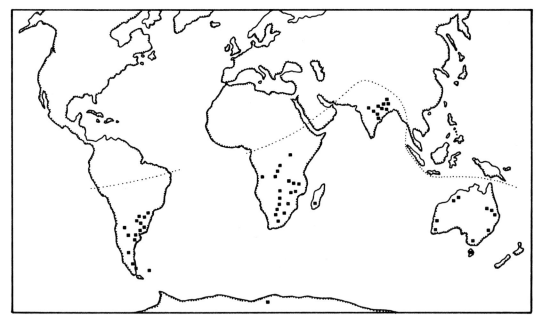
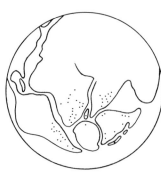

Distribution of the *Glossopteris* flora. On present day continental distribution it is scattered and disparate, but on the reconstruction of Gondwanaland (right) the same flora is brought into near continguity

material, volcanoes may achieve sufficient height to break above the sea as volcanic islands; a different kind of volcano may erupt along the line of plate consumption, as the plates plunge down and melt at depth.

The geophysical basis of plate movements is now being fully explored, although oddly enough it was the supposed lack of an appropriate physical mechanism that made the scientific establishment sceptical of the whole idea of continental drift fifty years ago. As far as the palaeontological world was concerned, the evidence of fossils was one of the first compelling bodies of facts to show that continental drift was a reality. For example, it was pointed out that the Carboniferous fossil plant floras with *Glossopteris* were remarkably similar in India, southern Africa and South America. Subsequently the same peculiar plants turned up in Antarctica. How could one explain such a distribution with the continents as they are at present? Certainly it seemed unlikely that these terrestrial organisms could have drifted across such wide barriers even if the climate were more uniform that it is today. And the idea of thin land bridges connecting widely separated areas seemed preposterous. However, if one fits the continents back together to form Gondwanaland (*see* above) the *Glossopteris* floras are brought into relatively close proximity, and none of these objections apply. The flora looks like a relatively continuous belt occupying what were cooler latitudes at the time. The same arguments were applied to terrestrial vertebrate fossils. Of course

animals and plants *are* capable of crossing bodies of water, but a look at the animals which have managed to reach remote islands today demonstrates the point that only a few kinds of animals succeed in doing so. The oceans are effective barriers.

The combination of faunal, floral, geological, and geophysical information leads inevitably to the conclusion that the present continents were fused together in one great 'supercontinent' in the Triassic. This is called *Pangaea*. Continental drift began in the Mesozoic to break up Pangaea into fragments – the present continents. These drifted apart slowly, with widening oceans between, a process which continues to the present day (*see* p. 33). This means that the basaltic sea floor of the Atlantic and Indian oceans must have been created at the ocean ridges since the Jurassic – it is all 'new' rock. And to prove this, no fossils older than Jurassic are known from the sediments which have accumulated in these regions, or indeed in *any* region of the sea floor. All the sea floors are, geologically speaking, young rocks, and the youngest of all are still being created in volcanic eruptions like the one that built Surtsey (off Iceland) in a matter of days.

In some cases the drifting continental blocks collided with one another. The most impressive results of this are mountain *chains*, like the Himalayas which were thrown up by the collision of the Indian subcontinent with the main mass of Asia. Such effects are highly dramatic, but it must be stressed again that the time over which it all happens is immense. The elevation of the Himalayas (which continues) has taken millions of years, and at no time would a hypothetical observer have been able to see the mountains bodily rearing from the sea as the great masses of India and Asia approached. The squeezing of the rocks along linear mountain chains results in their being folded, dislocated, and maltreated in many other ways which are a delight to a geologist, but sometimes a cause of gloom for the palaeontologist, because the fossils in the rocks are subjected to the same treatment, and may emerge at the other end much the worse for wear. The whole process of squeezing culminates in heating, metamorphism and sometimes melting of the rocks, which effectively destroys the fossil record, although there are some remarkable examples of fossils having survived enormous temperature elevation.

As Pangaea broke up, the dispersing fragments carried the fossil remains of the Permian–Triassic supercontinent to their present scattered positions. Conversely, the animals that lived on the drifting continents were then isolated, more or less, from their contemporaries in the rest of the world. For terrestrial vertebrates such isolation can result in the marooned animals having their own independent evolutionary history. Australia separated from Pangaea and carried eastwards a cargo of early marsupials which were isolated from further contact with the more advanced mammals that soon came to dominate the rest of the world. The isolation did not stop evolution, quite the reverse, for the marsupials evolved into a very varied group of animals. They were able to occupy almost all the niches that were available to them, from burrowing, tree climbing or grazing to carnivorous or scavenging habits. The primitive pouch is a common feature linking all of these animals, from the marsupial mouse to the great grey kangaroo. Some extinct giant marsupials died out after man first made his entrance into this self-sufficient world. The subsequent introduction of the domestic cat is doing a lot of damage to smaller marsupials; it does not take long to undo what tens of millions of years of continental drift have created. South America was similarly isolated until the recent geological past and another set of *endemic* mammals evolved, including the giant sloth (*Megatherium*) the bones of which Darwin collected in his voyage on the *Beagle*.

Continental drift has also had its effects on marine organisms, although they are more complex than for land animals. The free floating larvae of most marine organisms mean that oceans are not the barriers for sea creatures that they are for marsupials. Marine animals are, however, adapted to particular water temperatures, which is why the species of molluscs in the tropics are generally different from those found in the

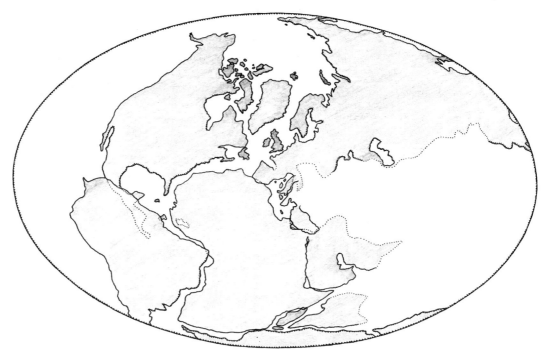

Reconstruction of the Permo-Triassic continent Pangaea, which has broken up to give our present continents

seas around the North Atlantic. The distribution of marine fossils can be used as a kind of thermometer to show how the water temperatures have changed as the continents, with their fringing seas, have changed position relative to the lines of latitude. In some cases the continents themselves act as a barrier for the marine animals. The general North–South direction of the Americas and Africa effectively isolates the Atlantic Ocean from the Indo-Pacific today, which has resulted in species endemic to either region. The opening of a seaway connecting these separate oceans quickly results in mixing of the faunas, a process which we have been able to see in action in the short time since the Panama Canal was opened.

In the last few years there have been attempts to trace the history of continental distribution back still further. Why should we suppose that plate movements only started with the disruption of Pangaea? More and more evidence is being accumulated that shows that Pangaea itself was only a phase in the development of the face of the earth. The supercontinent was the product of an earlier phase of drifting that bought together separate plates, so that in the Ordovician the world had a number of separate continents, as we do today. Our present continents were cut through in different ways, so that it is harder and harder to recognize the origins of the familiar continents as one goes back in time. If this process of drifting continued into the Precambrian then it becomes a very difficult business indeed to reconstruct the patterns of continents, and seas that have long since disappeared into subduction zones. The mysteries of these distant times are only now starting to be unravelled.

MAJOR CYCLES IN EARTH HISTORY: THE FLUCTUATING OCEANS

Fossils of marine animals can now be collected from rocks covering the far interior areas of continents which have been stable blocks since the Precambrian, regions which are now far removed from the oceans. In some cases marine deposits of this kind are explained by the fact that regions now isolated from the open ocean were formerly at the edge of continents when they were separate drifting plates. The Himalayas, the origin of which was described above, are an example where sediments that accumulated at the edge of the drifting Indian continent, and Asia, were sandwiched between the two

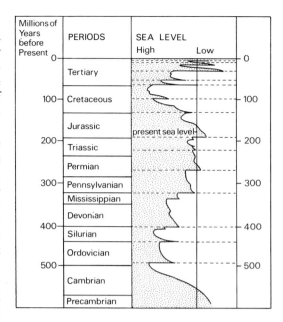

The fluctuations in the levels of the sea since the Cambrian (cycles of transgression and regression)

blocks as they collided. There is no problem in explaining the presence of thick marine deposits here, the former sites of seaboards that have been telescoped and crushed during continental collision. But in other cases, such as the interior of North America or Australia, there have not been comparable continental collisions and yet the sea evidently flooded over the continental interior from time to time, leaving deposits full of marine fossils extending deep into the heart of the continent. The only explanation of these kinds of deposits is that the sea has extended periodically much further over the continental areas than it does today.

These periods of flooding (or *marine transgressions*) are now recognized as one of the most important cycles of physical events that affected the earth, and their effects on the course of evolution have only recently been considered. Major marine transgressions of this kind have occurred at intervals since the Cambrian period (*see above*). Since they are due to the influence of rising sea level, they are simultaneous over the whole world. They afford another way of subdividing geological time. When the sea extended over the widest areas, naturally the deposits of that age are the most widespread and tend to be the most well known. Conversely, there were periods when the sea drained off the continents (*regressions*); at these times marine deposits will be confined to areas peripheral to the continents, and in the open oceans, while terrestrial sediments will extend over the areas where marine deposits were accumulating during the transgressive phase.

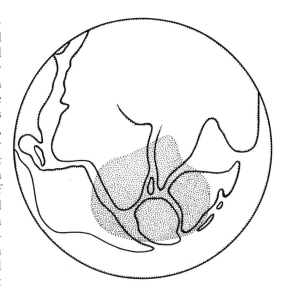

These oscillations may be related to continental drift. Some workers have suggested that the periods of transgression correspond with active phases of generation of new ocean floor at the mid-ocean ridges, when drifting was at its fastest, in fact. And the regressive phases correspond with periods of temporary standstill of drifting activity, when the continents stopped their slow progression for a while. Another cause that has been invoked are the Ice Ages. During a time of ice advance an enormous quantity of water is locked up in ice sheets, and world sea level falls as a result. As the ice melts sea level rises, producing a transgression. Rising and falling sea levels in the last million years were certainly controlled by glacial events. Whatever the cause, these great cycles have influenced the spread and evolution of life, both on land and in the sea. During the transgressive phase shallow marine faunas are widespread and diverse; in tropical waters reef bodies build up, and support some of the most diverse communities of marine animals.

MAJOR CYCLES IN EARTH HISTORY: THE FLUCTUATING CLIMATES

The continents move, and the seas advance upon them and drain off them again. This may already seem a world in which everything is in a state of flux, but to set the stage for the environment in which past organisms lived, we have to introduce one more cycle of change. The world climate has varied, from generally warmer at some times, changing to glacial phases at others. This has to be distinguished from changes of climate on any *one* continent: obviously if a continent is drifting it may pass across climatic belts, starting in the tropics and finishing up near the pole (like the southward drifting Antarctica), or vice versa. Climatic cycles are the amelioration or hardening of the climate over the earth as a whole. There is a connection with the sea level changes, as the major glacial episodes will probably result in a regressive (draining) cycle over the continents.

There have been four major periods of glacial activity in the period over which life has left its abundant traces in the rocks. The first of these was in the late Precambrian (there were almost certainly others in the further reaches of Precambrian time), just before the massive diversification of animals with hard parts at the base of the Cambrian. Another glacial phase was late in the Ordo-

vician. Probably the most important of all was the Carboniferous to Permian event, or rather events, whose effects are preserved over a huge area of the southern hemisphere (*see* above). The distribution of the glacial rocks here was used by Wegener in 1915 as one of the arguments in favour of continental drift. At present, Africa and Antarctica span a great range of latitudes on today's continental distribution. On the predrift map they slot neatly together, just what might be expected from a polar ice-cap. The effects of the glaciation in the southern hemisphere were felt in Carboniferous Europe, which at that time lay near the equator. Glacial retreats released immense quantities of water, which resulted in sea level rises. On some occasions the rise was sufficient to 'drown' the coal measure swamps, bringing the sea over what had formerly been luxuriant jungle, and interpolating marine fossils in rock successions which are otherwise full of fossil plants, insects, and amphibians. The Upper Palaeozoic glaciation lasted for a long time; between the major pulses in the Carboniferous and the Middle Permian there was an interval of perhaps 20 million years.

We are currently still within a glacial phase. The Pleistocene period, extending back over the last million years, is the latest of the glacial episodes. Over this period, radiometric dating techniques, and other geophysical criteria, combined with careful analyses of fossil pollens and various microfossils, has enabled a subdivision of geological time on a much finer scale than for the earlier parts of the column. The story so revealed is very complex. The ice has advanced and retreated numerous times; the

Glacial 'pavement' with the scratch marks made by an over-riding ice sheet. Such marks can be found repeatedly where there are sites of former glacial activity

original idea of four great advances has been shown to be a gross underestimate. The number of glacial phases is now thought to be fifteen or so. Phases of retreat between advances included times when the climate in Britain, for example, was very much milder than it is today. There is no reason to suppose that the process is at an end; another ice advance may be on the way in the future. The huge quantities of water locked in the present ice sheets mean that the area of exposed continent is probably greater now than was general in the Tertiary and before.

Between glacial phases there were other periods in which the climate of the world, at least as recorded in the rocks, was generally warmer and more equable. The Silurian and much of the Cretaceous period were times of unusual general warmth. At these times the tropical zones expanded to cover a greater area of the oceans and continents. Wide spreads of warm-water limestones are typical of the rocks deposited at such times, and with the limestones are found fossil animals adapted to the warm oceans.

Paradoxically, there are probably more species of animals *in toto* at times when there was a gradient of climates from cold at the poles to warm equatorial. Each climatic zone permits its own suite of adapted faunas. A uniform, warm climate results in widespread faunas, but rich in species though these faunas may be, they may not be as rich as the sum of the faunas of all the climatic zones, at times when there were ice caps.

THE CHANGING WORLD

The arrangement of the continents change, the sea levels change, and the world climate changes. This is the shifting stage on which organic evolution acted. The cast of characters changed repeatedly; the animals or plants were adapted to the conditions pertaining while they lived. In this dynamic world, what are we to make of Hutton's ambition to interpret the geological past by processes that operate at the present day? The basic assumption still stands, that though the stage may have changed, the processes that shaped the scenery were the same in the past as they are at the present. Physical laws do not change. The sediments deposited in the Permian deserts extended over different parts of the globe. It is only by a thorough understanding of the forces at work today that the past can be reconstructed. Life was bound up with the story of the changing earth, and it is foolish to pretend that the history of life can be fully understood without its dynamic setting. The history of life has been so closely bound up with the history of our planet that it is likely that some small change in that history would have produced a change in the course of evolution. If the climate had not changed in Africa a few million years ago, would Man himself have evolved? There is nothing inexorable about the course of evolution; rather it is a complex *pas de deux* between the changing environment, and the capacity of organisms to respond to those changes.

ROCKS AND FOSSILS

A visitor to a museum will see perfectly preserved and spectacular fossils neatly arranged in glass cases. It is easy to forget that the fossils were found by cracking open rocks. All fossils are found in rocks that were originally unconsolidated sediments. The study of fossils can be enhanced by knowing about the rocks that enclose the fossils. Certain environments which today support a rich and varied plant and animal life have no sediments forming in them, and the organisms living there have virtually no chance of being preserved in the fossil record. Mountainous regions, for example, are dominated by the erosion of the rock forming the ranges, and therefore no permanent sediment is formed there. Torrential rain and rapid weathering, aided in some climates by the action of frost, rapidly destroys much of the organic material: the chances of any preservable remains reaching a lowland river where permanent sediment is accumulating are remote. The faunas and floras of mountainous regions of the past are most unlikely to be represented in the fossil record. The *fossilization potential* of a mountainous environment is low.

The study of fossils is connected with a suite of rocks that are formed in environments where sediments accumulate, and have a high chance of becoming rocks. Such environments cover a large part of the surface of the globe, including most of the submarine areas, and some of the lowlands, where the rivers and lakes accumulate sediments of many types. This chapter is concerned with the different sites in which sediments may form.

In a general way the site of sediment accumulation is also a direct reflection of the environment in which an animal or plant

lived. Lake fishes and plants, for example, are to be expected in *lacustrine* (deposited in a lake) sediments. The wide variety of sediment sites will be reflected in an equal variety of animals from the different habitats. Sediments accumulating in different sites have distinctive characteristics which are found in those from no other environment. Extensive studies of recent sedimentary types enable the interpretation of sediments from the past. Most sedimentary rocks retain in fine detail the features acquired while they were accumulating. So by studying recent sediments it is possible to determine the site of deposition of past rocks and from this to understand more about the environment in which the fossil animals in the rock lived.

Occasionally dead animals and plants travel for long distances before finally becoming entombed in the sediment. Empty shells of *Nautilus* have been found over a much wider area than that in which the animal lives. Drifting logs can be found hundreds of miles from land, and when these become waterlogged, they sink and eventually become incorporated into the sediment.

SEDIMENTARY FACIES
The rocks formed in a particular site, each with their own peculiar characteristics, are called *sedimentary facies*. Just as at the present sediments of many different facies are accumulating in different places, so in the past rocks with totally different appearances may have accumulated at the same time in different environments. The fossils found in such rocks also differ from one site to another, because they were also related to the environment in which the sediment accumu-

lated; different sedimentary facies may have different assemblages of fossils. The term *facies fauna* is then applied to an assemblage of different fossils that are found together in one particular sediment type. Not all animals are so limited: in the sea, for example, many free swimming or floating organisms are independent of the conditions of sediment accumulation on the sea bottom.

The diagram below shows the main sedimentary facies in a hypothetical section running from the mountainous interior of a continent in tropical latitudes to the open ocean.

In the lee of the mountain range a rain-starved desert accumulates mostly wind-blown sand derived from weathering of steep buttes. Occasional torrential bursts of rain produce flash floods, which sweep down the steep walled valleys (or wadis) carrying with them all the weathered material, which spreads out into broad fans. The flood water drains into temporary pools, which evaporate rapidly in the hot sun, but sometimes linger long enough to allow brief bursts of specialized animals to colonize the warm water. As the waters evaporate, mineral salts are concentrated within the pools, often crystallizing out as the drying tends to produce white, glistening salt pans. Most of the animal and plant life here has to be strictly specialized to cope with the harsh conditions.

On the mountains themselves erosion predominates, aided by the action of ice found on the higher peaks. Frost shattering helps to splinter rocks into shards and blocks that tumble down slopes and become the raw material for streams to move inevitably downwards towards the sea. Melting snow produces raging torrents with immense transporting power, that in full spate can move and break huge blocks into smaller cobbles which are more easily transported. Little sediment accumulates here except in deep depressions between ranges (inter-montane basins), where lakes may form, the scree slopes at the edge of the mountains, and in the deposits of streams and rivers. Much of the sediment produced on the mountain ranges becomes ground into smaller particles. In this state it can be transported by the large and more sluggish rivers through the foothills and beyond. Occasional floods originating in the mountains may transport huge amounts of material, often with disastrous consequences for flooding on the plain.

The deposits associated with major rivers are varied silts, clays and sands, often with characteristic combinations of sedimentary structures that reveal their *fluviatile* origins. Large lakes lying on the plains are another important and potentially fossil-bearing site of sediment accumulation. In low lying areas swamps support prolific vegetation which decays to form beds of peat. Insects and other animals adapted to this habitat may be destined to be preserved therein.

As the rivers wind towards the sea their flood plains broaden, and they meander over the plains, breaking their banks and

Sedimentary environments. A section from a desert area through a mountain range, across a sedimentary plain to the shallow sea, and ultimately to the deep sea, showing the different sites in which fossils may accumulate.

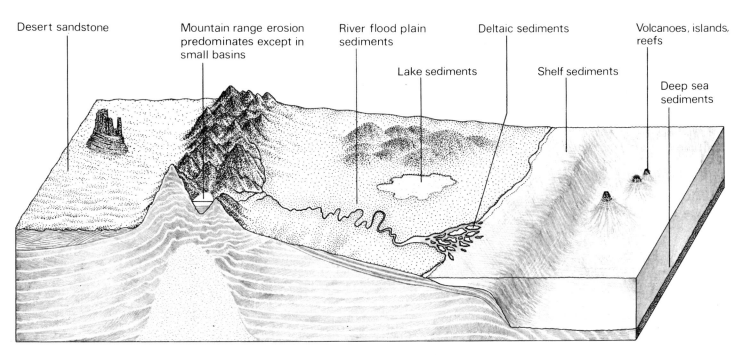

Desert sandstone

Mountain range erosion predominates except in small basins

River flood plain sediments

Lake sediments

Deltaic sediments

Shelf sediments

Volcanoes, islands, reefs

Deep sea sediments

readjusting their courses at time of flood. They still carry huge quantities of sediment, mostly in suspension (at the present day about 10,000 million tonnes of sediment finds its way into the seas over the whole world) but also in solution. At the junction of land and sea the river begins to shed its load, partly from the effect of fresh water meeting salt, and partly because its energy and carrying power dwindles. Typically the river builds a delta, and the silts, sands and clays of deltas are one of the most important sedimentary facies in the fossil record. Swamps may form about the small streams (distributaries) that criss-cross the delta, and there are many habitats suitable for the successful life of animals and plants, and their rapid preservation on death. Deltas slowly build out into the sea, forming an advancing wedge of sediment that may extend for many miles. In such cases the seaward edge of the delta is younger than its fossilized edges preserved landward in the deltaic sediments. So the delta sediments, although in geographic continuity, are not everywhere of the same age. These are *diachronous* deposits.

Smaller rivers, like the Thames, often do not build deltas but enter the sea in estuaries, where salt and fresh water oscillate in influence in tidal stretches of water. Estuaries also produce characteristic sediments, and have specific sets of organisms associated with them. They are not in general as geologically significant as deltas.

We now pass into the marine environment, which has produced the bulk of the sedimentary rocks and in which the record of past life is most complete. In many areas the contact between land and sea is erosional, as any visit during a storm to a resort with sea cliffs will prove. Storm beaches of rounded and tough cobbles often form the boundary between the sea and the land, and may be preserved in the rocks as coarse *conglomerates*. More offshore sediments are generally finer sands, silts and clays, and far more suitable for preserving fossil remains than the deposits of storm beaches. The finer material from rivers and from the direct erosion of the land by the sea contributes to the sediment deposited on the sea bed. Quite powerful currents on the sea floor often influence the pattern of distribution of such sediments. Currents are necessary to bring food to many of the inhabitants of the shallower sea floors that will become preserved as fossils. Even areas that look quite barren, like sand flats, can support quite a variety of life living within burrows under the surface of the sand and feeding only when covered by water. Shallow marine deposits again have characteristic sedimentary features. The ripple marks of tidal seas (*see* below) are frequently encountered in rocks, and the tracks of animals can be preserved here. Fossil footprints in sand are no doubt preserved somewhere today, recording the holiday habits of mankind. Although there are many exceptions, the grain size of the sediment generally decreases away from the coast, so that the more offshore deposits tend to be fine muds that will produce clays or mudstones in the geological column.

Recent ripples (*left*) and fossil ripple marks preserved in sandstone (*right*). The detailed structure of the ripple marks can reveal much about the sedimentary environments

An oolitic limestone, showing the perfectly rounded ooliths of which it is composed. Such limestones formed in shallow-water, agitated marine conditions, and only in warm climates. This example is Jurassic in age.

Globigerina ooze; a sample from the deep sea largely composed of the minute tests of planktonic foraminiferans. Twenty times natural size

In warm climates, where there is not much land-derived sediment available, other forms of sedimentation become important, especially the formation of limestones. Many limestones are formed from the consolidation of lime muds. The lime in these muds is a form of calcium carbonate (aragonite) in impalpably minute crystals, which can only be thrown down from solution in warm and shallow seas. Where there is turbulent agitation of the water in shallow areas the carbonate may be laid down in concentric layers around tiny organic nuclei to produce rounded ooliths. A rock largely composed of ooliths is an *oolite* (*see* above). In spite of their rather specialized mode of formation oolites can form a surprising volume of limestone formations, covering hundreds of square miles in the Ordovician of North America, and almost as extensive in the Carboniferous and Jurassic rocks of Europe.

Numerous species of animals are confined to a lime substrate, and naturally one finds their extinct counterparts preserved in limestones. Some limestones are composed very largely of the remains of calcareous animals; these rock-building fossils will be considered in more detail in a following section.

As the deep sea is approached there is generally a reduction in the amount of sediment that can reach the open ocean from land sources. On the floor of the open ocean at great depths large areas are covered with a fine ooze, which is formed predominantly from the skeletons of minute planktonic organisms that have rained down from the surface waters. The greatest density of life exists in shallow depths where light penetrates and allows microscopic plants to live, and there are small planktonic organisms which feed on these plants; it is the shells of these small animals that form much of the deep sea sediment. The rate of accumulation of these deep sea deposits is very slow, only a few mm per thousand years. Several kinds of single-celled organisms may dominate these deep sea oozes; the foraminiferan *Globigerina* (*see* p. 41), which has a calcareous test, and the delicate, siliceous radiolarians (*see* below) are especially important. These deep sea deposits record a remarkably complete history of the evolution of the planktonic organisms forming them. Even these small shells are to some extent soluble in sea water, and at very great depths pressure increases this solubility so that they do not survive. Here the only deposit is the *red clay* (often in fact brown in colour), a sediment which accumulates extremely slowly. It is composed of the finest wind-born dust, volcanic ash carried by winds from distant eruptions, and, occasionally, the insoluble traces of ocean-going animals such as the teeth of sharks or the ear bones of whales (*see* p. 43). In these *abyssal* depths curious nodules with a high proportion of manganese grow slowly in the red clay, and there has recently been speculation on the possibility of exploiting these as a mineral resource, surely the least accessible ore in the world. In spite of the inhospitable, lightless conditions in the abyssal seas there is a variety of life, and there are specialized and often bizarre fish and crustacea that live only there. These leave little fossil record.

In the open ocean, volcanic islands form sporadic sediment sources, both from the erosion and redistribution of the volcanic rocks themselves, and because they reproduce the same sort of conditions that pertain on the continental shelves. The pure, clear water surrounding such islands in the tropics is often suitable for the growth of coral reefs.

In Arctic regions (*see* p. 43) the influence of ice as both an erosional and depositional agent is paramount. The scouring action of ice, using rocks enclosed within glaciers as tools to scrape and gouge the underlying rock surfaces, produces great quantities of angular detritus. Some of the rock is ground literally to a flour. At the melting edges of glaciers, or where icebergs break off from ice caps and drift into the sea, much of the material is released and becomes sediment.

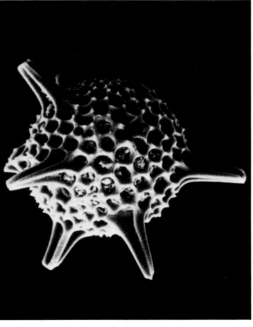

The delicate, siliceous skeletons of radiolarians cover large spreads of the deep sea floor. About 300 times natural size

The ear bone of the whale

Sedimentation in Arctic regions. The influence of ice is paramount, both dumping sediment at the edges of ice sheets, and rafting it into the open ocean

sources of their bones. Further away from the ice front major rivers took away the melt waters to the sea, and their *fluviatile* deposits often preserve the remains of the large mammals that lived in the surrounding areas, some no doubt fatally entrapped in bogs.

Some land-derived sediments reach the deep sea by means of turbidity currents. These are slurries of sedimentary material that are flushed from shallower areas at the edge of the continental shelf, a movement often sparked off by earthquakes. Sometimes their effects can be quite catastrophic, snapping cables and the like, and they can travel extraordinary distances, up to 200 miles or more. Turbidity currents produce a characteristic rock type in the geological record (*see* p. 44), known as a *turbidite*. Communities of animals which live on the bottom and suddenly have material deposited on top of them by a turbidity current can be both killed and buried in a single catastrophe.

Just as climate influences the kinds of sediment laid down, so also it is one of the most important influences on animal and plant life. Few species are truly worldwide. Most marine organisms are zoned according to latitude, and it is possible to represent these distributions as a series of belts approximately parallel to lines of latitude (distorted by the influence of warm and cold currents) (*see* p. 45). The same sort of influences undoubtedly operated in the past. During the Pleistocene, when climates oscillated over many thousands of years between warm and cold, marine and land organisms migrated backwards and forwards with the

Such glacial deposits (*till*) are often a heterogeneous selection of different rock types, dumped together, with large boulders and tiny pebbles immersed alike in the groundmass of rock flour. Not surprisingly fossils are rare in these kinds of rocks. But around the edges of the ice sheets mossy bogs are common and may form depostis of peat and lignite containing the remains of the organisms adapted to life in high latitudes. During the last Ice Age animals often used caves as shelters or lairs, and the deposits of cave floors have proved particularly rich

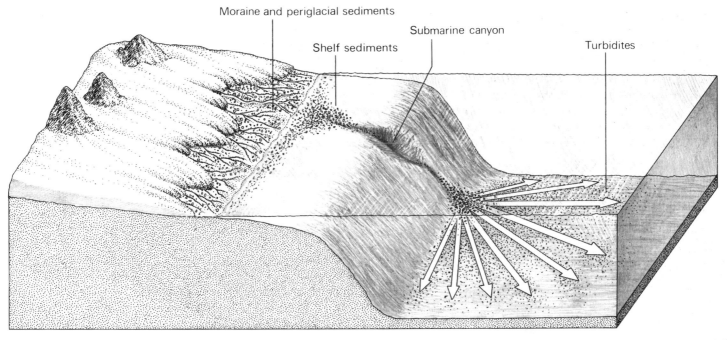

Moraine and periglacial sediments

Shelf sediments

Submarine canyon

Turbidites

43

An Ordovician turbidite – the fossil deposit of a turbidity current. The coarser material is at the bottom of the bed. This rock is on a beach where it has been turned vertically – like a groin. Some of the beach gravel is visible at the top of the picture.

– Fine division

– Cross-bedding

– Coarse division

climatic shifts to keep living in the conditions to which they were adapted. Since these oscillations ran approximately parallel in the land and in the sea, this provides one of the methods of subdividing the Ice Ages. Tropical faunas and floras are richest in numbers of species, and, at the other extreme, only a few hardy species of high Arctic or Antarctic animals and plants are able to cope with the extreme conditions. Those species that do adapt to Arctic conditions may be found in great profusion.

We have seen how sediment type alters with the depth of water beneath the sea. Animals are also influenced by water depth, and it is not surprising to find different communities of fossils in rocks that were deposited at different depths. In their different ways both the former sediments and the animals found as fossils within them are reflections of the environment, and both provide clues to the life conditions of former times.

REEFS

Organic reefs today are formed by corals and calcareous algae that make wave resistant structures, often of immense dimensions. Because the framework of a reef is composed of tough calcareous organisms which have to stand up to the buffeting of waves and the ravages of storms they are highly likely to be preserved in the fossil record. Reefs can be preserved with all the constituent organisms in life position. Living reefs can only be found in warm water regions (23–5°C is best) with abundant sunlight. Light is needed by the minute photo-synthetic algae that live in the tissues of some reef-building corals. Reef growth can only proceed satisfactorily in salt water, and not in water of too great a depth where the vital light begins to be filtered out. Nor can they tolerate too much clogging sediment in the surrounding waters. The majority of fossil reefs seemed to have lived under similar constraints.

The reef environment is an extremely rich one, supporting a host of animal species besides the frame builders themselves: fish, molluscs, sea urchins and crustacea that live among the corals as scavengers and predators, together with numerous encrusting organisms that benefit from the firm foothold provided by dead coral. The corals themselves feed by filtering minute planktonic food from the constant wash over their polyps. Erosion of the reefs is rapid, and this builds slopes of reef waste and finer

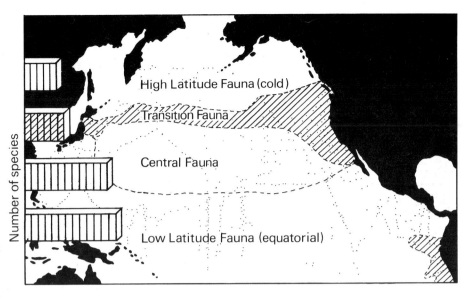

debris that can also be found fossilized. This produces a rock that is almost entirely made of fossils. Darwin was the first to demonstrate that the curious circular pattern of atoll reefs (below) was due to the foundering of volcanic islands, around which reefs had initially formed as fringing structures. The rapid growth of the reef can keep pace with the sinking of the island, producing in

Climatic zonation of the living planktonic foraminifera

Darwin's theory of the formation of coral atolls; as the volcanic island sinks the reef begins to assume its typical circular pattern

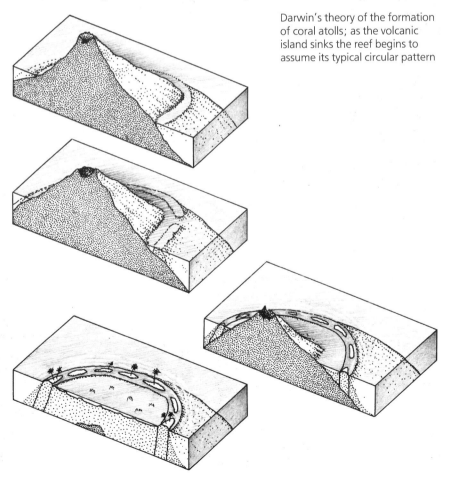

the process a great amount of sedimentary waste that preserves the earlier, fossil history of the reef.

The reef as a structure has a long history, extending back even to the time before the corals themselves had evolved. Other organisms were able to adopt the same role as the corals, often assuming similar general shapes, even though unrelated in the zoological sense. The earliest reef structures are already present in Lower Cambrian rocks in Labrador, where the frame is built by enigmatic sponge-like organisms called archaeocyathids. Corals themselves, although belonging to forms unrelated to living species, started to achieve prominence in the later Ordovician, and by the Devonian large reef structures were formed, although the storm-facing surface of the reefs was often formed by massive stromatoporoids, a group of organisms with only a few inconspicuous relatives living today. Carboniferous reefs are almost worldwide. Some of the corals from these reefs closely resemble their living counterparts (*see* below) although this resemblance is due to similar life habits, and their detailed structure is quite different.

In the Permian rocks of Texas and adjoining States large reef structures have been formed with sponges and bryozoans as the important frame builders, along with the ubiquitous algae. The same habitat supported one of the most bizarre brachiopods, a conical form with a 'lid' quite unlike the usual lamp shells (*see* Chapter 4). At the end of the Permian most of the reef-building organisms that built large reefs in the Palaeozoic became extinct. Triassic sponge reefs are known. By Jurassic times coral reefs were again being constructed, this time by the distant relatives of the corals and other organisms that build reefs at present. The corals of Palaeozoic age are distantly related, if at all, to those of the Mesozoic to Recent, but they both built reefs of similar construction. During the Cretaceous another extraordinary group of organisms

Carboniferous reef coral (*left*) which assumes a very similar general shape to the biologically unrelated species on the right, which is from a recent reef

acquired the habit of building reefs. These were the rudists (*see* above), which again are conical, with a lid, at first glance much like the aberrant brachiopods of the Permian. However, the rudists were quite unrelated to the brachiopods, and were derived from clam-like molluscan ancestors. The rudists did not survive the Cretaceous, and were confined to very warm limestone seas; they may have been adapted to higher temperatures than pertained over much of geological history. Even though Upper Cretaceous limestones are common in Britain, rudists are rare fossils here, but they are important rock formers in the Alps and in North America.

During the Tertiary, the break up of the continents, with the associated volcanic activity creating islands in the ocean, permitted the establishment of the ancestral reefs that continued to be built until the present.

The reef environment is important because it shows how different organisms can assume similar superficial appearance when they adopt similar life habits. There is more than a passing resemblance between some of the archaeocyathids of the Cambrian, the Permian richthofeniid brachiopods and the Cretaceous rudists, and some of the corals have comparable form as well (*see* p. 46). On a reef itself the more densely branched forms today tend to be found on the more exposed, seaward flanks, while the backreef, protected areas often have organisms with loosely branching antler-like growths. There may be variation in branching habit even within a single species, according to the site in which it flourishes. Through geological time various organisms have played the same ecological role, and the result has been similar shapes. This is an ecological control that can act on very different starting material (like bivalves and brachiopods) and produce a superficially similar end-product. The biological prerequisite of most of the organisms mentioned here is that they should be filter-feeders. Conversely, for a highly adapted organism like a reef dweller it is necessary to look carefully at the fine structure to determine the biological affinities of the organism, and not be misled by superficial resemblance.

DEEP SEA DEPOSITS

The peculiar sediments formed in the deep sea, beyond the edge of the continental shelf, also have a long history. Deposits laid down in the open ocean occupy a large area of the globe today, and there is every reason to suppose that this was true far back in the Palaeozoic. Yet the areas occupied by oceanic sediments in areas of Palaeozoic rocks is not commensurate with this former extent. This is because most of the ancient oceanic sediments have been destroyed where they plunge down the subduction zones at the consuming edges of plates; oceanic sediments are the ones that disappear forever. The crust of the earth is in a state of dynamic equilibrium, with the oceanic crust being created in some oceans at the mid-oceanic ridges, only to be consumed at the edges of others. The break up of the supercontinent Pangaea during the Mesozoic resulted in both the creation of the oceanic crust that floors the oceans today, and the destruction of the previous oceanic crust, so that at present the oldest ocean crust is probably only Jurassic in age.

In spite of the odds against it, some of the ancient oceanic deposits *are* preserved, but only in special circumstances. Sequences of turbidites form prisms of sediment, sometimes thousands of feet thick, at the edges of the continents. During phases of continent closure these sediments become squeezed between the approaching continental plates as the oceanic crust itself is consumed. The sediments respond to the pressure by variously crumpling, shattering, and gliding into great sheets that move away from the centre of pressure. Many of these sediments become heated, partially melted, or so contorted by pressure that any fossils they once contained are transformed beyond recognition. But others survive with their fossils intact, although the strata from which they have to be recovered are almost always vertical, rather than in their original horizontal attitude, and frequently grievously distorted, with the fossils they contain distorting along with the rock. The fossils are not usually found in the turbidites themselves, but in interbedded shales, representing the quiescent conditions between turbidite slumps. In some cases slices of the true bed of the open ocean, instead of being consumed, have been thrust beyond the danger zone, carrying on their backs a skin of sediment. These slices have a characteristic combination of volcanic rocks (often with serpentine) with cherts and sometimes black shales. These remnants of former oceans are known as *ophiolites*. Where ophiolites are found in mountain belts this is taken as indicating that an ocean has been consumed in that region, and that adjacent areas of folded or metamorphosed rocks were produced by continent colliding with continent, the intervening ocean which was originally present having dived to oblivion.

In order to find deposits laid down in the ancient deep seas and the fossils they contain, the right geological setting must be found. It is of no use looking in the sediments formed on the ancient Shield areas of Precambrian rocks, which even in the Palaeozoic were covered by relatively shallow seas. But in the areas of intensely folded rocks which probably accumulated at the edge of former continents, the chances of finding deep sea sediments are increased.

Even in the Cambrian the deep sea shales between turbidite sequences have been found to contain trilobites – minute, blind forms, known as agnostids. The likelihood is that these specialized trilobites were free swimming or planktonic forms, and that they did not actually live on the inhospitable sea bottom in the abyss. In the Ordovician and Silurian drifting colonial organisms known as graptolites usually dominate any assemblage of fossils recovered from between the turbidites, their remains often forming matted clots (*see* p. 49). Clumps of graptolites, perhaps killed at the surface by a plankton 'bloom' consuming the available oxygen, slowly fell through the water column, eventually to lie down on the soft muddy surface; the arrival of another turbidite flush made their entombment complete. The occasional glass sponge may have actually lived on the deep bottom; they are found sporadically through the record in these sorts of sediments, and are still relatively prolific at the bottom of the oceans today. Special extraction techniques can usually recover the remains of single celled plants from deep sea deposits of Palaeozoic (or even Precambrian) age. These show that photosynthetic algal plankton, the basic link in the food chain of the open ocean, was present from the earliest times. No doubt the seas also swarmed with minute zooplankton feeding off the tinier plants, some of these being the larvae of the more familiar fossils. Such zooplankton is mostly soft-bodied and leaves hardly any record, but the graptolites may well have been the next link in the food chain. The carapaces of some extinct crustaceans are known in some abundance in graptolitic deposits; they may have fulfilled the same function as the open sea shrimps today, but there was certainly no Lower Palaeozoic 'whale' to harvest them. Cherts accompanying the ophiolites often contain the remains of microscopic radiolaria, which had acquired their siliceous skeleton even by the Cambrian, and must have had planktonic habits then, as at the present.

The graptolites became extinct early in the Devonian, and no ocean-going animals of this colonial type are found subsequently as fossils. At about the same time the early relatives of the ammonites were adapting to life in the open seas, and their coiled shells, variously ornamented, become the fossils most frequently encountered in deep-sea sediments for several hundreds of millions of years. Ammonites swarmed in vast num-

Clots of graptolites of only one or two species can be preserved in deep sea deposits in the Lower Palaeozoic

bers both over the seas of the continental shelves and in the open sea, and the total mass of living matter they must have represented over their geological life span from the Devonian to the Cretaceous is almost inconceivable. Some species may have swum in masses, moving gregariously like their distant relatives the squid, which occur in huge numbers in the open seas today. By the later Palaeozoic fierce predators added another level to the food chain, the sharks by then being well-advanced, and doubtless as voracious as they are in modern seas. The sharks are a primitive group of fish in their structure and organization, but are superbly well-adapted to their role as killers, which has ensured their survival into modern times, when their early prey species, such as ammonites, have passed into extinction. Along with ammonites other kinds of invertebrates, like delicate clams, sometimes occur in some profusion. These may either have been free swimming forms, or attached to floating seaweed. The micro-fossils include an increasing variety of species of planktonic algae which by the Jurassic included forms related to those still living. The radiolaria continue their unbroken his-

Ammonite 'lids' or aptychi have been recovered from supposed deep sea deposits. Twice natural size

tory from the earliest fossil-bearing rocks. In the Jurassic the foraminifera, hitherto almost exclusively bottom-living forms, took to life in the surface waters of the ocean, and played an increasing role as sediment makers, which continues until the present day. During the Cretaceous minute calcareous platelets which coated the outside of algae (coccoliths) (see p. 160) are an important component of pelagic sediments, for all their small size showing a wonderful variation in symmetry. The slow build-up in variety of animals contributing to oceanic sediments continued to the late Cretaceous, when, apparently quite rapidly, the ammonites became extinct.

Whatever the cause of the demise of the ammonites it did not affect the foraminifera to quite the same extent, although the post-Cretaceous ones are all new forms. For

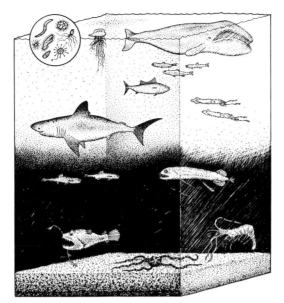

Depth zonation in open ocean animals; the sea floor receives their dead remains regardless of the depth at which the animal may have originally lived. Not to scale

were inimical to virtually all forms of life. No doubt these crisis periods were implicated in the changes that occurred in the composition of oceanic fauna and flora However, it can be shown that the changes that occurred on land, for example among the mammals in the early Tertiary, did take place more rapidly than those detectable among the planktonic foraminifera. But the cumulative and inexorable changes have resulted in a complete transformation of oceanic life over 600 million years.

NICHES NOT REFLECTED IN THE ROCKS

The sediment surface indiscriminately receives the scraps that will become fossils, and it would be a mistake to assume that any animal when alive necessarily lived where its remains finish up. And any one environment is usually subdivided into innumerable micro-habitats. The variety of animals living together is explained by the number of *niches* (particular ways of earning a livelihood) into which a broader environment can be divided. Many of the details of the niche are not reflected directly in the rocks. The details of the life habits of the fossil plants or animals then have to be inferred from different evidence. Then there are cases where an animal's livelihood depends on the numerous kinds of organisms which, being soft bodied, have left virtually no record. Polychaete worms are an abundant source of food in shallow marine environments, but their only geological legacy can be slight disturbances of the sediment produced by their burrowing activities.

Among living planktonic organisms there is often a stratification according to water depth; most species live near the surface, but there are some that live at deeper levels in the water column. The sediment surface below receives both kinds with equal ease when they die. Depth stratification of this kind has been suggested for fossil foraminifera and graptolites. In this case the record of the rocks can be used to test the theory (*see* above). Shallow water planktonic species will have a distribution through much of the world in the surface waters, but the deep species will be restricted to areas where the water depth is sufficient. The deepest basins should have the richest assemblages of species in the sediments, because it is in such areas that all the depth zones will be stacked up above the sea floor.

Cretaceous and younger rocks we can investigate the deposits of the deep seas directly by taking cores from the bottom of the sea, or looking at rocks where they have been uplifted on volcanic islands. During the Tertiary the oceans soon acquired a modern aspect, with fish, whales, crustaceans of modern type, and micro-organisms related to those of present seas. Nonetheless a succession of different species can be recognized, with certain types becoming extinct, and others appearing as 'landmarks' for the divisions of the last 70 million years.

The sheer volume of the oceans makes the oceanic environment relatively stable compared with the terrestrial one. Extreme changes of temperature, for example, are muffled in the sea, and below a certain depth the temperature is virtually the same the world over. Particularly in the open ocean the animals are cushioned from the effects of environmental fluctuations and variations which have provoked the rapid evolution of terrestrial organisms. Comfort usually only stimulates laziness, but we have seen that the oceanic environment has, in fact, changed considerably since the time in which it can be first certainly recognized in the rocks. The radiolaria and planktonic algae have been with us all that time, but have not remained static, and successful groups, like the graptolites, have failed entirely, while the planktonic foraminifera have successfully adapted to the challenges of the open ocean. It is now known that there were certain periods when the oceans did undergo crisis. These times often entailed the increase and spread of sea water lacking oxygen, and

CHAPTER FOUR

HOW TO RECOGNIZE FOSSILS

Fossils are not rare. We have seen how millions of fossils may make up the rocks themselves, crammed together layer after layer to form formations thousands of feet thick. The present land surface is a thin skin on top of a thick record of the past preserved in the rocks. Once scientists began to collect the record of past life they soon came across the problem of how to order and arrange the huge variety of fossil forms they recovered. In order to understand what they had found they needed to classify the fossil organisms into particular kinds, more or less similar, to impose an order on what would otherwise have been a vast and chaotic mass of different and apparently unrelated relics.

The urge to order the natural world appears to be an innate characteristic of human beings: 'a place for everything and everything in its place'. But classifying animals and plants, living and fossil, is more than just an attempt to find a convenient way of slotting them into different categories, like stamps in a stamp album, for neatness and convenience. The classification of the natural world is supposed to reflect the great ordering *process* that itself gave rise to the variety and diversity of animals and plants that are alive today: the process of evolution. Animals that look similar are classified together, and not only that, they are also closely related in an evolutionary sense (or put another way, they share a common ancestor). So when we look at the grimacing gestures of a chimpanzee and wonder at the almost ludicrous parallels with our own behaviour this is just part of a whole host of behavioural and anatomical similarities that show without doubt that we ought to be classified with the apes (we are all of us primates), and that we share a distant an-

cestor with our diminutive caricatures. Evolution has driven man and the chimpanzee further apart from our shared ancestral species, and in evolutionary terms the chimpanzee is as advanced as we are, although man dominates by virtue of numbers and adaptability. Few people are offended today, as they were in the last century, by the thought of man and the chimpanzee being classified together by virtue of a common ancestor which has been extinct for probably more than three million years! But enough family likeness remains to make the chimpanzees' tea-time a most popular attraction in zoos all over the world.

A word of caution here: not *all* similarities indicate relationships with evolutionary significance. Some similarities can be misleading, because animals can superficially resemble one another that are not closely related in an evolutionary sense. Often these resemblances are the product of a similar mode of life. Both tortoises and armadillos are animals that carry around their own suit of armour, and at first glance we might think they were related. But a little further investigation shows that tortoises are cold-blooded reptiles that lay eggs, while armadillos are in many respects typical warm-blooded mammals, bearing their young alive. Obviously tortoises and armadillos cannot be classified together in spite of their similarities. When classifying fossils even more care is necessary because we have not got as much evidence to go on as with living animals. We cannot see directly whether a fossil animal was warm or cold-blooded when alive; often the evidence is indirect and depends on careful study of little bits of bone. The palaeontologist is like a detective trying to reconstruct a full story from a few

Distantly related species may come to resemble one another closely. Slow worm (*left*), snake (*right*) and amphisbaenid (*bottom*), probably all descended from different ancestors, and reached their similar shapes along separate evolutionary pathways.

fragmentary clues. He must be careful not to follow any 'red herrings' that will result in him classifying his fossil on the basis of ambiguous similarities. He must distinguish snakes from eels, tortoises from armadillos, on the basis of the bones presented to him. To do this, he must have thorough familiarity with the living fauna, because the information on living animals is so much more complete. We shall see in the next chapter how carrying comparisons with living animals *too* far can result in curious and inaccurate pictures of the past.

The number of species of animals and plants living today runs into millions, and similar numbers of species have probably lived on earth for at least part of the earth's history. Perhaps it is fortunate that the fossil record preserved only a fraction of the truly stupendous total number of species that must have lived since the Cambrian, for otherwise the scientists' task to catalogue 600 million years of life would be an impossible one. It is not surprising that new fossil species are discovered daily, and indeed the amateur collector has a good chance of finding a new species of fossil, if he looks hard enough and learns to recognize what he has found.

There must have been an increase in the number of different kinds of animals and plants since the Precambrian; for example, the conquering of land alone gave rise to a multitude of new opportunities for the colonizing organisms, resulting in an increase in the total number of species. Within the marine environment itself some palaeontologists believe that the overall richness of the marine fauna was established early on, say by the Silurian period, and that there has not been a great increase in the total number of species living in the sea at any one time since then, although of course the *kinds* of organisms inhabiting the sea have changed many times. And there have been several major episodes of extinction (Chapter 6), when the cast of characters in the sea changed almost totally. Although the number of species may have been at least approximately the same in marine environments for the last 300 million years or so the kinds of fossils have changed repeatedly, so that, for example, in marine limestones of Silurian age the shelled brachiopods may number dozens of species, whereas in similar looking limestones of Eocene age no brachiopods at all can be found, but there may be as many species of gastropods of kinds unknown in Silurian rocks.

Even by the beginning of the Cambrian period, when fossils start to become easy to find and many different kinds of animals had acquired preservable hard parts, it is possible to classify the fossils found in the rocks into broadly similar groups. Nearly all of these groups correspond with major divisions of living animals, so the broad base of classification was established by the Cambrian. These major groups continued to evolve throughout the following 600 million years, and they include the distant ancestors of our living fauna. Not all the major living groups of organisms were present in the Cambrian, however, because the colonization of the land did not take place for another 200 million years, and so there were obviously no direct ancestors of land plants around, or of most of the land animals with which we are familiar today. The major groups into which we can fit almost all fossils from the Cambrian onwards, and all the living fauna, are called *phyla* (singular: *phylum*). To identify any animal properly, the first stage is to determine to which phylum it belongs. Of course even the phyla themselves originated from unknown ancestors, and they all ultimately derived from the first living cell. Fossil evidence for some of this early history is shrouded in obscurity, hidden in the vast stretches of Precambrian time, but as is shown in Chapter 7 more evidence is being discovered each year. Some of the answers are only now coming to light as we find out more about the structures of the proteins that go to make up living cells themselves. By the time organisms had become sophisticated enough to have hard parts, evolution had already defined the phyla, and the great natural framework for classification had been almost completely constructed.

At the opposite end of the scale from the phylum is the smallest unit of classification usually used for fossils: the *species*. To be strictly accurate a species can be defined precisely only in living animals, where it refers to populations that can interbreed under natural conditions, and which produce offspring that are capable of further reproducing their kind. In practice, many species are recognized by having some peculiarity of shape, behaviour, plumage, colour and so on, that reliably set them apart from all other similar species. Obviously we shall never know whether fossil species were capable of interbreeding until somebody invents a time machine, and we can go back and see for ourselves! So our fossil species are defined in a rather practical way, as showing some consistent difference or differences from all other related species. Because of the patchy nature of the fossil record many species (particularly of large vertebrate animals) are known from only one specimen; some of these are of almost inestimable value, which is why museums have to take good care of their material. Every species has a unique scientific name. The species are 'christened' when a scientist describes them for the first time, illustrates their peculiarities and publishes the name in a scientific journal.

The species name has two words: *Tyrannosaurus rex* is a familiar example. The names are usually derived from Latin or Greek – '*Tyrannosaurus*' means 'tyrant lizard' and '*rex*' 'king' so the animal is appropriately described as King of the tyrant lizards: *rex* is the *specific name*, unique to this species. There may be other dinosaurs, which are similar to *Tyrannosaurus rex*, but belong to another species. These may be included in the same *genus* – *Tyrannosaurus* – but will be given a different specific name. Some species are named after their collectors, so if our new species of *Tyrannosaurus* were collected by a Mr Jones it might eventually be christened *Tyrannosaurus jonesi*. Incidentally, the generic name always starts with a capital letter, and the specific name with a small one, even if it is named after Jones.

In this chapter typical examples of the kinds of fossils most commonly encountered are illustrated by beautiful specimens. Very few fossils have everyday names, and so the scientific name is used. Once a few have been mastered it is surprising how quickly the most ponderous sounding scientific name acquires a familiar ring. The scientific name has the great advantage of being the same all over the world, so the language of nomenclature is a truly international one.

Between the species on the one hand and the phylum on the other there are a whole series of intermediate categories of increasing inclusiveness. Several genera (plural of genus) may be grouped together in a *family*, several families together in a *superfamily*, and the superfamilies themselves are clustered into *classes*, two or more of which combine to make the phylum. It all sounds rather complicated, but it does serve a useful function in ordering the almost countless number of species in the most economical way. All the species within any of the units of classification descended from a single, ancestral species. We need not worry about families or superfamilies in this book, but classes often correspond to everyday groups of animals. In the phylum Mollusca – the molluscs which include the majority of sea animals with shells – the class Gastropoda includes all the snails, and the class Bivalvia all the clams (mussels, scallops, etc.). Many of the most familiar fossils, like the trilobites, are classes within larger phyla.

Plants are classified in much the same way as animals, but just to make things difficult the largest units of classification are not generally called phyla but 'divisions'. Many of the plants familiar in the garden – Dahlia, Chrysanthemum, Fuchsia – are generic names which have come into common language. The fossil record of plants is rather patchy compared with that of some animal phyla. Partly because of this the broad level classification of plants is still being argued.

We will now examine the commonest and most important kinds of fossils, the kinds that the reader will be able to find when he starts a collection. The groups of organisms described have certain special peculiarities that set them apart from all others, and once these features are recognized the group to which the fossil belongs can be confidently identified. Each group is illustrated in the plates by a typical species, but of course all the groups contain many species, and any specimen the reader is likely to find will probably differ from the one chosen for illustration in several details.

Most accounts of the different kinds of fossils start with the simplest organisms and work towards the most complex, often ending somewhat selfishly with man. This can give the impression of a simple evolutionary tree: single cells to many cells, shellfish to fish, and then land vertebrates. This is not correct. Although single-celled organisms are obviously simpler than complex animals like ourselves, many single-celled organisms have continued to evolve actively since they lived in the 'primaeval soup'. It is known that some of the unicellular organisms in the sea today are only one or two million years old (very little in geological terms), in fact they are probably as young as man himself! Very simple and very complex organisms have lived side by side for a long time and both have evolved together. True 'living fossils' are really rather rare, and the term can be applied to both simple and highly complex organisms that have outlived the time when the earth was populated with many more of their kind.

It is easier to describe simple organisms first and move on to complex ones but this is just a convenience of arrangement, and we will follow this procedure here. Some of the most important fossils are extremely tiny: these will be discussed in a later chapter, and what follows is concerned with the forms that can be recognized from hand specimens. When they are first chipped out of the rock such fossils will often be partly concealed by matrix. To identify them it is usually essential to clean off most of the enclosing rock, otherwise one can be misled by superficial resemblance. The rule is never to try such cleaning in the field – it nearly always results in a pile of rubble and a frayed temper.

THE DIFFERENT KINDS OF FOSSILS AND HOW TO RECOGNIZE THEM

SINGLE-CELLED ANIMALS – PROTOZOA
(Colour plate 2)

A large and heterogeneous collection of organisms are grouped in the Protozoa, including all those animals that can lead an autonomous existence as a single cell. Most are microscopic, and many have no skeleton and therefore lack a fossil record. The flexible amoeba, which is the protoplasmic 'blob' of popular imagination, is a familiar protozoan without much potential for fossilization. A few protozoan groups secrete hard *tests* (shells) which are very common fossils, but again most of these are too small to be easily spotted in the field. They are of great geological importance, however, and we shall return to them later in this book. Such testate protozoans have a record extending back to the Cambrian. A very few protozoans secrete skeletons of (for a single cell) gigantic size, and these also occur in such numbers that they form conspicuous and common fossils. Particularly in rocks of Eocene age the coin-like *nummulites* are important as rock builders. Nummulites are known from England, but they are most numerous in the Mediterranean and eastwards, where they form limestone formations of great thickness. The pyramids of Egypt are constructed of limestone blocks in which the species *Nummulites gizehensis* is conspicuous (opposite p. 64). Nummulites are a giant kind of foraminiferan, an important group of rock-forming organisms as far back as the Carboniferous (and with ancestors in Cambrian rocks), and which still form deep sea oozes today. Most foraminiferans require a microscope for their study. Another group of giant foraminiferans flourished particularly in the Permian period: these are the *fusulines*. Instead of being about the size and shape of a coin, the fusulines are spindle-shaped, with a round cross-section tapering at both ends; fusuline limestones again form thick rock sequences, particularly in Russia and the Orient. Both fusulines and nummulites are divided into many small chambers internally, and details of their internal structure are used to classify them. Both groups seem to have flourished in exceptionally warm, shallow seas in the tropics of the Tethyan region.

THE SPONGES – PHYLUM PORIFERA (Colour plates 1, 3)

The sponges are an important group of many-celled organisms, with the individual cells specialized for particular functions but rather loosely aggregated. Some sponges have the property of being able to 'rebuild' their colonies of cells if they are mashed up. Marine sponges have a long fossil record from the Cambrian onwards, and at many localities they are abundant enough to be important rock formers. They have been prominent reef-building organisms too, often in association with bryozoans. Not all sponges contain hard parts capable of being fossilized. The skeletal elements of those with hard parts are fine, often branching elements called *spicules*. The spicules may be only loosely associated within the sponge tissues, or may be more or less fused to form rigid skeletons. Two major divisions of sponges are distinguished on the composition of the spicules: the glass sponges have a skeleton composed of silica, one of the few examples in nature where this material is used as a basis for a skeleton, while the calcareous sponges use calcium carbonate for their spicules. Both kinds have Cambrian representatives and both flourish today, the glass sponges being especially

numerous in deep-sea environments. The sponge skeleton is highly variable in shape: many are cup or flask shaped, others spherical or cauliflower-like, while some form flat plates folded together, or are encrusting. A successful group of sponges has taken to boring into the shells of other organisms. In detail, sponges are distinguished by the form of the spicular structure, and the ways these combine to form the skeleton, as well as the overall shape. Although sponges are widely distributed in the fossil record they are particularly numerous and easy to collect in Cretaceous rocks in Europe and North America, where dozens of well-preserved species have been described.

Sponges are often preserved inside flints, which were originally deposited as hard layers within the Cretaceous chalk. As they are much more resistant to weathering than the enclosing chalk, the flints remain behind when the chalk is eroded away. Flints containing sponges may be incorporated into younger sediments (they are then known as *derived fossils*), and it is not unusual for such flints to be dug up in suburban gardens around London – a long journey from the Cretaceous seas.

CORALS AND RELATED ANIMALS – PHYLUM COELENTERATA (Colour plates 4, 5, 6, 9)

The coelenterates are a varied group of organisms with cells organized into definite

The calcareous sponge *Raphidonema faringdonense* from the Cretaceous of Berkshire, England. This sponge with a calcite skeleton was obtained from a gravel, an unusual type of rock in which to find fossils. It is a broadly cup-shaped sponge, rapidly expanding from a narrow base. The minutely porous external surface is covered with irregular bumps. A number of species of this genus have been reported from Cretaceous rocks, and related genera occur back to the Triassic. Species related to this one occur in France, Germany, Italy and Switzerland. Sponges vary a good deal in size and shape. The specimen shown has a height of 7 cm. Specimens from unconsolidated gravels are not difficult to clean, any adherent sand grains being easily removed using a stout pin.

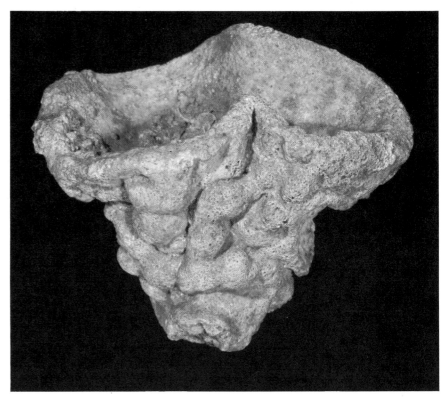

tissues, a mouth surrounded by tentacles, and a stomach. The tentacles are arranged in a circlet about the mouth, and the coelenterates are typified by such *radial* symmetry. The majority of coelenterates have a jelly-like consistency, and the group includes the familiar jellyfish and sea anemones, as well as the smaller hydroids commonly kept in schoolroom fresh water tanks. Coelenterates come in two main designs: they are either attached with the tentacles surrounding the mouth waving like an animal flower in the water (*polyp*), or free floating with the tentacles hanging down, often as fine as hair (*medusa*). Coelenterates have stinging cells which they use to paralyse their prey. It is surprising to find that the soft-bodied jellyfish have any fossil record at all, but in fact they have the longest one of the phylum. Casts of the stomach cavities have been described from Precambrian rocks in many parts of the world, and many different Precambrian jellyfish have now been named. It

The lithistid sponge *Pemmatites arcticus* from the Permian of the arctic island of Spitzbergen (*left*). The sponge skeleton has weathered out of a limestone. Unweathered specimens are sometimes difficult to see on the rock surfaces. This sponge superficially resembles a living bath sponge more than the other sponges in this book, being bun-shaped, with a minutely pitted external surface. *Pemmatites* species are known from a number of localities in the Permian and Carboniferous of Europe. This specimen has a diameter of 7 cm. To obtain a hand specimen we rely on the differential weathering of the sponge and the matrix. Some related forms are better studied in polished section like the corals.

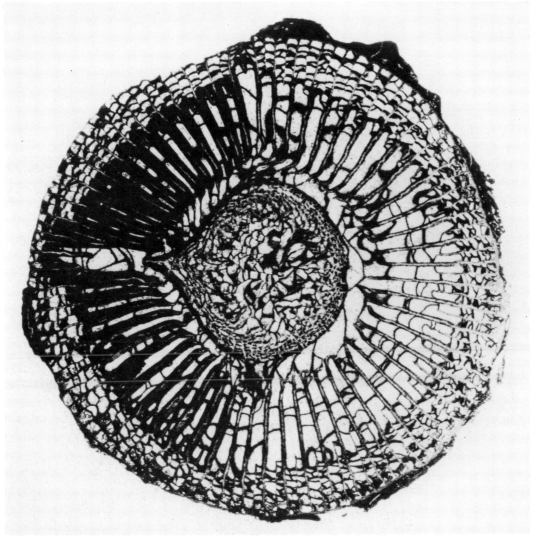

Section across a typical rugose coral (*Palaeosmilia*, Carboniferous)

Devonian corals, *Heliophyllum corniculum*. These silicified specimens have been weathered out of the rock. They are small, simple corals with a gently curved form. Note the deep cup which was occupied by the *polyp* (the living coral animal) in life. Corals with this simple form are known from Ordovician to Carboniferous rocks, but there are differences in detailed structure. *Heliophyllum* has numerous species in the Devonian of Europe and North America. Length of the whole coral is about 3 cm.

is certain that the seas of 600 to 1000 million years ago swarmed with drifting medusae, just as they do at present. These free-floating coelenterates are therefore both primitive and ancient. Some groups within the coelenterates have developed the ability to secrete a hard skeleton, of which by far the most important are the corals (Class Anthozoa).

A coral is essentially a sea anemone which supports its body by a skeleton of calcium carbonate. This first happened early in the Ordovician (some believe even earlier), and there were no doubt simple polyps in the Cambrian and Precambrian from which the corals evolved. The secretion of a skeleton gave the corals a great advantage over many other coelenterates; supported thus they could overcome the lack of rigidity of their soft tissues and grow large. Many coral species form branching colonies in which hundreds of individual polyps live on top of their own houses. Other forms remain *solitary* – a single polyp. From soon after their inception colonial corals started to live together in large masses with one or many species forming large, mound-like, wave resistant structures – coral reefs. Many organisms other than corals also contribute to the construction of the reef. Fossil reefs are one of the prime sites to look for fossils of many different kinds besides the corals themselves.

Corals can be divided into several major groups. In the Palaeozoic, corals looking generally similar to living reef corals may be only distantly related, if at all, to our present day fauna. These *rugose* corals can be solitary, or massive in large reefs; in either case their skeletons are composed of the form of calcium carbonate known as *calcite*. The finest details of the skeletal structure are well-preserved, and the corals can be cut and polished, or studied in thin sections (*see* p. 57). The detailed structure of the individual plates that combine together to build the skeleton are the basis for the classification of the corals, as well as the general form. The most obvious of these building elements are plates arranged radially, looking like the spokes of a bicycle wheel in section (*septa*). In rugose corals the septa usually have a basic four-rayed symmetry in each corallite. Rugose corals are common fossils in rocks of Ordovician to Permian age.

Alongside the rugose corals low cushions or branching masses of a different kind of calcite coral are often found. These have much smaller corallites usually only a millimetre or two across, and the septa typical of rugose corals are absent or inconspicuous. These are the *tabulate* corals (Colour plate 5). Both rugose and tabulate corals are extinct.

Corals looking superficially very like the rugose ones are found in Mesozoic and younger rocks, and are important on reefs today (*scleractinian* corals). Their calcium carbonate skeletons are composed of the mineral *aragonite*, chemically the same as calcite, but with the atoms arranged in a different way. The septa in these corals are often arranged about a six-fold symmetry so that they are fundamentally different from the Rugosa. Oddly enough, although they are younger than the rugose corals, the aragonite composing them does not preserve very well, and it is easier to find beautifully preserved examples of the older Rugosa. Scleractinian corals are found in small numbers in Triassic rocks, and are abundant in Jurassic, Cretaceous and younger rocks, particularly in past and present tropical regions.

THE GRAPTOLITES – PHYLUM HEMICHORDATA (Colour plates 7, 10)

The graptolites are an extinct group of colonial organisms, with a geological record extending from the Cambrian to the Carboniferous. For many years they were regarded as colonial coelenterates, but it is now certain that they are unrelated to the jellyfish and their allies, and in fact are distant cousins to a small group of tube-dwelling organisms with little fossil record, which belong to the minor phylum Hemichordata. Hemichordates are a primitive group that probably share a common ancestor with the chordates (including vertebrates). The graptolites swarmed in the seas of the Ordovician and Silurian periods. They are usually preserved as flattened impressions, which retain little of their finer detail. The impressions generally show a serrated edge like a saw, each 'tooth' being the crushed tube (*theca*) that housed an individual of the colony (*zooid*). The colonies vary widely in shape: some are shrub-like, with numerous slender branches, others have only a few, or even a single branch. The bushy ones (Order Dendroidea – dendroid graptolites) are the more primitive, and were generally rooted to the sea floor. The few-branched forms, which include the Order Graptoloidea or planktonic graptoloids, were derived from the dendroids at the base of the Ordovician and were successful and prolific until the early Devonian. They then disappeared, to be survived by their more primitive bushy relatives. The number of branches and the arrangement of

Jurassic reef coral, *Stylina alveolata*. This massive coral is now composed of calcite. It has been recrystallized, because in life it was composed of the mineral aragonite – chemically the same as calcite, but with a different crystal structure. The matrix is a yellow limestone common in the European Jurassic. Rounded, massive coral composed of individual corallites with about ½ cm diameter. The septa that line the edges of the corallites are here very short. The little cups occupied by the polyps in life have been exposed by the weathering. There are a number of species related to this one in coral reefs and clumps of Jurassic and later age. The specimen is from Nattheim, Germany; related species of *Stylina* occur in Jurassic limestones in Europe, eastwards into Asia Minor, and in the US. Diameter of the entire colony is about 10 cm.

Eight-branched graptolite, *Dichograptus octobrachiatus*, Ordovician. This graptolite has eight equal branches, each lined with many tiny tubes in which the individual animals of the colony lived, although most of the microscopical details have been destroyed. *Dichograptus* species are very widespread in the early Ordovician rocks. Some species have very slender branches. This species is found in the eastern United States, Texas, Britain, Canada and Australia. The diameter of the colony is 6 cm.

the thecae are important in their identification.

The graptoloids are one of the important groups used in dating rocks. They evolved rapidly and spread widely, and with a little experience a glance at an assemblage of graptolites on a shale surface can be used to determine the approximate age of deposition of the rock. Most of the later graptolites had only a few branches and, in the Silurian, species with only a single branch tend to dominate the assemblages. One advantage of graptolites is that they occur predominantly in former oceanic environments – the deeper water shales or limestones discussed in Chapter 3. Their preservation in inner shelf habitats is much more unusual – although not unknown. Whether the graptolites were not living nearshore, or whether the conditions were not right for their preservation, are questions that have excited some argument. Whatever the explanation, other organisms, like trilobites or brachiopods, were more numerous in such sites, and have been used to date the rock sequences in the absence of graptolites.

BRACHIOPODS – PHYLUM BRACHIOPODA
(Colour plates 8, 11, 12)

A first collection of marine fossils will almost inevitably include a brachiopod or two. Brachiopods have one of the longest histories and one of the best fossil records of any invertebrate. They are already present in early Cambrian rocks and are still with us today, although living brachiopods tend to be rather inconspicuous in shallow waters. But during the Palaeozoic and Mesozoic they occurred in such profusion in inshore sediments that they are frequently important components of the rocks in which they are found.

Brachiopods have two valves, and this gives them a superficial resemblance to bivalved molluscs. The drawing below shows that bivalves and brachiopods really differ in fundamental symmetry, and the differences in the shell reflect even more profound ones in the internal, soft anatomy. Brachiopods have probably always filter-fed, living off small organic particles brought on currents. The particles are harvested by a *lophophore* covered in cilia – which serves the dual purpose of both creating a current and catching the food. Most brachiopods were attached to the substrate by a stalk, and the hole through which the stalk entered the shell can often be seen on the fossils. This method of food-gathering and the inactive mode of life may seem sufficiently dull for us to anticipate that the brachiopods would have changed little in their long history. It is true that one brachiopod – *Lingula* – is one of the most famous 'living fossils'. *Lingula*-like species, looking much like the living form, can be found in rocks as old as Ordovician. But the other brachiopods have been far from evolutionarily inactive – they have gone through several major proliferations and diversifications, and suffered dramatic major extinctions as well.

In the Ordovician and Silurian periods they became adapted to life in most marine environments, but were particularly numerous in shallow water habitats, in some cases forming whole banks, as mussels do today. Although generally small (and hence easily collected) some species grew to 10 cm long or more. Some brachiopods are smooth, but

Brachiopod (*left*) and bivalve mollusc (*right*) compared, showing difference in symmetry when viewed sideways, reflecting fundamental internal distinctions

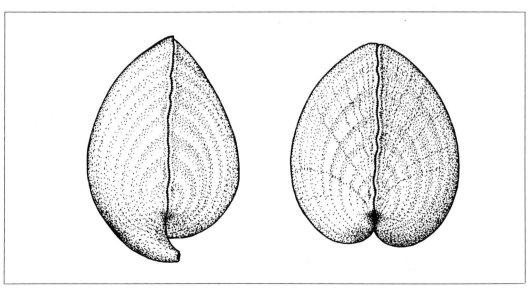

many became corrugated and ornamented with coarse or fine ribs. The margins of the valves are often wavy, and deeply folded in other species. Long spines on the exterior of the shell were developed especially during the Carboniferous. Even more profound changes happened in the internal structures of the brachiopods, particularly those concerned with supporting the lophophore – these changed from simple loops to complex 'doubled back' structures, or to fantastic spirals and whorls – all presumably designed to increase the ways of food-gathering, and its efficiency. Quite complicated tooth-and-socket hinges were developed between the valves. Although the variety of brachiopods in the Jurassic and Cretaceous is somewhat less than in the Palaeozoic, they are still very abundant and varied fossils. It has been suggested that their slow decline in the last 100 million years or so has been caused by the commensurate rise in diversity of filter-feeding bivalve molluscs, which ousted them from their former habitats. It is an attractive idea but one difficult to prove; in any case many of the greatest successes of the bivalves have been in life habits that the brachiopods *never* adopted (burrowing and swimming free, for example).

Brachiopods fall into two major classes; the more primitive have shells composed of calcium phosphate plus organic material and hinge development is imperfect – hence their name, inarticulate brachiopods. In-articulates can be recognized on the rock from their shiny lustre. *Lingula* is one of the best known genera. Inarticulates were numerous in the Cambrian and Ordovician periods. The other class, the hinged brachiopods (Articulata) have shells composed of calcium carbonate, and they comprise a more varied and abundant group as fossils. Through their long history from the early Cambrian to the present different groups of articulate brachiopods rose to prominence only to decline. As with other invertebrates the combination of brachiopod types in an assemblage gives a quick clue to the age. Modern studies rely particularly on the internal structures for their identification: some brachiopods with similar exteriors can have very different 'insides'. But the overall shape of some kinds is distinctive enough for quick recognition – the large spiriferid brachiopod shown on Plate 11 is unlikely to be confused with any other, and brachiopods of this type are abundant in the Carboniferous limestone. Above all it is the sheer variety and abundance of brachiopods that give shallow water Palaeozoic assemblages of fossils their distinctive 'feel'.

The brachiopod *Macrocoelia expansa* is a common species on this slab of Ordovician age. These brachiopods are preserved in a distinctive way. The rock surface is covered with both internal and external moulds. No trace of the original shell material remains. The enclosing matrix is a fine-grained sandstone, and the original calcite shell material has dissolved away to leave impressions of the shells on the sandstone. This sort of preservation is very common in European Ordovician and Silurian rocks. The fine radial ribbing is characteristic of many such brachiopods; internal moulds show the impressions of strong hinge teeth (showing as hollows on the rock surface). Species like these are abundant in rocks of Ordovician age, often forming shelly bands. This species is from Near Meifod, Wales, but similar species are very widespread. Precise identifications are difficult, and require expert knowledge. The rock specimen is 20 cm across.

Jurassic spiriferid brachiopod, *Spiriferina walcotti*. This elegant brachiopod is characterized by its broad shell, thrown into folds at its margin. The hinge line is wide. Faint concentric growth lines can also be seen. The species shown here is from the Lias rocks of England and its close relatives may be found worldwide. It does not exceed 3 cm in length.

Fossil lamp shell, *Terebratula maxima*, Pliocene of East Anglia, England. The specimen shown has both its original valves preserved, and looks today much as it did when it had just died. This *Terebratula* is a particularly large brachiopod, as its Latin name 'maxima' implies. One of its valves is larger than the other, and that valve has a circular opening at its apex, through which a fleshy stalk passed which served to attach the animal during life. It lacks the strong ribbing seen on many brachiopods. Species superficially resembling *Terebratula maxima* may be found all over the world in rocks going back at least to the Jurassic period. Its close relatives, in the Tertiary rocks, are rather rare. Adult shells are 10 cm or more long.

SEA MATS – PHYLUM BRYOZOA (Col. plate 13)

Bryozoans are an important group of colonial organisms which form encrusting mats on other marine shells or rocks, or branching, leaf-like or hummocky colonies on the scale of a few centimetres. Their colonies can often be seen on seaweeds stranded on the shore, where they resemble a fine, white net attached to the darker weed. They are rather common fossils from the Ordovician onwards, and in spite of their small size can be abundant enough to be important as rock-formers (especially in limestones). Bryozoans found as fossils have calcium carbonate (calcite) skeletons that have a high chance of being preserved, and the fossil forms seem to have been exclusively marine (some living bryozoans live in

fresh water). A microscope is essential for their proper study, but even with a hand lens a glance at the surface of a bryozoan colony reveals a number of tiny openings. Each one of these openings was, in life, occupied by an individual of the colony (*zooid*), a minute animal with tentacles covered in cilia, that entrapped passing micro-organisms and edible particles. Larger colonies would have had hundreds of individual zooids. In more advanced bryozoans of the Mesozoic to Recent individual zooids have become specialized for particular functions – some have become totally modified to curious, snapping structures looking remarkably like miniaturized parrot heads, which may function to prevent the settling of larvae of unwanted alien organisms on the bryozoan colony. The zooids themselves are, or course, not preserved as fossils – we only have their vacated homes. But from a study of detailed thin sections through the colonies we can deduce a good deal about the growth (*astogeny*) of colonies of bryozoans long since extinct. Although they were filter-feeders like the graptolites the bryozoans have not taken to free-floating, planktonic existence – they are characteristic benthic organisms. They are divided into a number of groups on the basis of details of structure of the individual 'boxes' that housed the zooids, and the construction of the colony. It is always worth examining the surfaces of fossils like brachiopods or sea urchins to see if the fine matted or delicately branching colonies of bryozoa are preserved on their surfaces – they are easily overlooked. Other bryozoan colonies are more immediately conspicuous, particularly the stout, twig-like branches of the Palaeozoic trepostomes, which can make up thick limestone beds, and formed their own 'reefs', or the large, often net-like colonies of the 'fenestrellids' common in the Upper Palaeozoic. A relative of the latter, the curious, screw-like *Archimedes* is shown on p. 65.

Bryozoans are quite commonly preserved as internal moulds – the calcite skeleton is dissolved away leaving only the sediment fillings of the chambers once occupied by the zooids. In this case they can present a different appearance, a host of little tubes combining together into branching twigs or nets. Such preservation is usual in sandy or silty rocks. The variety of colony structure in the bryozoans exceeds even that in the corals, and ranges from compact humps to cobweb-like branching colonies of excep-

tional delicacy, and their small size in no way detracts from the beauty and variety of form they display.

MOLLUSCS – PHYLUM MOLLUSCA

The molluscs are one of the most varied, successful and numerous of the invertebrate phyla. Thousands of living species occupy every marine habitat from the shallowest shores to the deepest abyss, and, as every gardener knows who has tried to protect his vegetables against marauding hordes of snails and slugs, they have been remarkably successful in making the transition from sea to land. The most primitive molluscs lack a shell, but the most diverse groups, and those that concern us here, have well-developed hard parts that are readily fossilized. Molluscs have the lower part of the body developed as a muscular *foot*, which may variously function in locomotion, digging or swimming. The molluscs were ultimately derived from a soft-bodied creature somewhat like the living flatworms, but the acquisition of hard parts happened very early on, low in the Cambrian or before, and by the Ordovician the important living classes were well established. A few molluscs, like the slugs and octopuses, have secondarily lost their shells, having developed other means of protection. As with the other phyla, the great span of geological time has seen different groups rise to prominence, decline, to be replaced by others. But unlike the brachiopods the molluscs are probably as diverse today as they have ever been. Measured by their total living weight (biomass) molluscs are one of the most important groups in the whole marine biosphere: the swarms of squid in the oceans are the match of any species of fish. The level of organization achieved by the most advanced molluscs, for example the octopus, is the most intricate and sophisticated of any invertebrate: one cannot over-estimate the importance of the Mollusca in shaping the patterns of marine communities we see today. Different molluscs fill most of the possible ecological roles available to marine organisms: some are voracious hunters, others graze on algae or feed directly on organic material in muddy sediments, others again are filter-feeders. The great majority of molluscs have a minute, planktonic larval stage, a small ciliated object bearing no resemblance to the adult, that drifts as part of the plankton until ready to settle and assume its mature form. Thus even bottom-dwelling molluscs can be dispersed widely

1 Fossil glass sponge, *Hydnoceras tuberosum* from the Devonian of New York. This exceptionally large, beautiful sponge is preserved in full relief in a fine sandstone. *Hydnoceras* is a Devonian and Carboniferous genus, but glass sponges of similar general form (but without knobs) have a history going back to the Cambrian, and surviving species today. A large species with a length of 20 cm.

2 The giant foraminiferan *Nummulites gizehensis* (nummulite) from the Eocene of Egypt. Numerous specimens of this species are preserved together, actually forming the rock – an example of fossils as rock builders. The small, lentil-shaped fossils are a separate form of the same species. The 'coins' are up to 4 cm across.

3 The hexactinellid sponge *Coeloptychium agaricoides* from the Cretaceous of Westphalia, Germany. These two specimens are beautifully preserved, extracted from a matrix of white chalk, which gives them their colour. Species of *Coeloptychium* can be found in England and elsewhere in Europe. The 'cap' of this specimen is 8 cm across.

4 The solitary rugose coral *Cyathophyllum* sp., Devonian of Devon, England. Sections have been cut through well-preserved coral in limestone. The gaps between the walls of the coral skeleton have been filled with calcite, showing up as the lighter colours of the sections. The longer diameter of the section is 6 cm. Corals are often best sectioned like this one to reveal the internal structures.

1

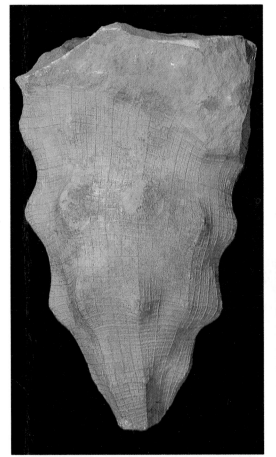

2

3

4

5 The tabulate coral, *Halysites escharoides* (chain coral) from the Silurian of Ohio Falls, USA. This massive coral has weathered out of limestone because the material of the coral is slightly harder than the enclosing matrix. Large numbers of tiny oval corallites are linked together to form chains. Specimen 10 cm across, part of a larger colony.

6 Fossil brain coral, *Coeloria labyrinthiformis* (Scleractinia). Miocene, Antigua. Polished section. Long meandering corallites with thin septa form a dense meshwork to give this coral an appearance like the labyrinth implied in the specific name. Specimen 8 cm across.

7 Dendroid graptolite. *Dictyonema flabelliforme*, Ordovician, North Wales (European geologists often regard rocks with *Dictyonema* as latest Cambrian). The large colony of this net graptolite is preserved in a light-coloured shale. Of the many species of *Dictyonema* only a few have a colony as regular in shape as this one, which can grow to a length of more than 20 cm.

8 Lingulid brachiopod, *Lingula beani*, Jurassic, Yorkshire, England. These tongue-shaped brachiopods are composed of calcium phosphate, with a shiny lustre, contrasting with the clay matrix.

9 Pliocence coral, *Septastraea forbesi* Maryland, USA. A massive coral with irregular knobs, composed of corallites with a diameter less than ½ cm, with polygonal outlines, and with about 12 prominent septa like the spokes of a wheel. The specimen is 10 cm long, and is a fragment of a much larger colony.

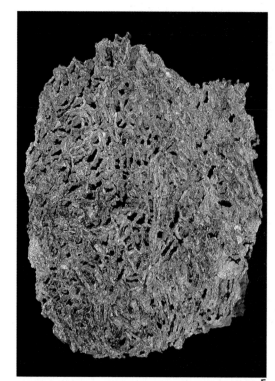

5

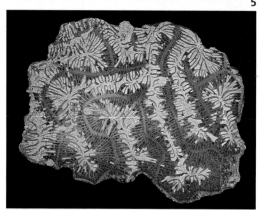

6

7

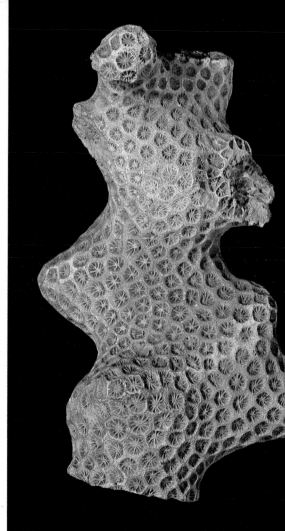

9

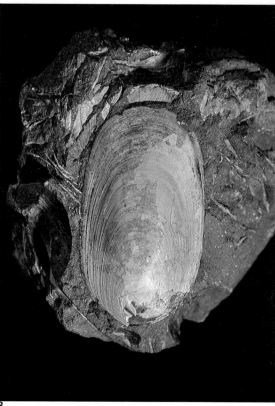

8

10 Tuning-fork graptolite *Didymograptus murchisoni*, Ordovician, Wales. Several of these graptolites are preserved on the flat bedding surfaces of a black shale. The graptolites are flattened, and their original skeletal material has been destroyed. *Didymograptus* species of this type have a distinctive shape like a tuning fork. Individual specimens grow to a length of 5 cm or more.

11 Brachiopod, *Spirifer striatus*, Carboniferous, Kildare, Ireland. This brachiopod has a wide hinge line, and the apertural margin is deflected downwards to form a broad 'v'. The radial ribs are more numerous than those of the related brachiopod *Spiriferina* p. 63. These specimens are 5 cm across.

12 The rhynchonellid brachiopod *Cyclothyris difformis*, from the Cretaceous of Devon, England. Wide, ribbed shells with a small *beak* projecting from the upper valve, the lower valve is deeply convex. *Cyclothyris* is found in both Europe and North America. These specimens are 3–3½ cm long.

10

11 12

13 14

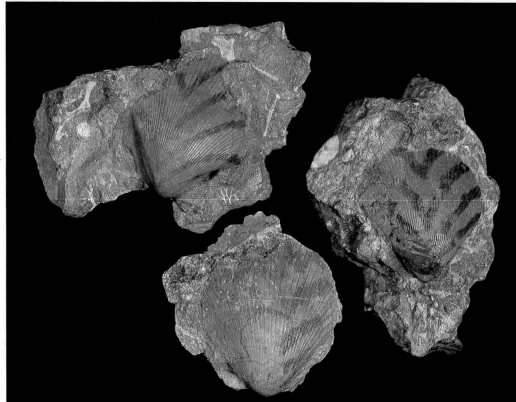

15　　　　**16 17**

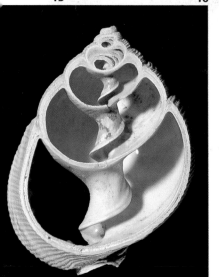

13 Net bryozoan, *Fenestrellina plebeia*, Carboniferous of North Wales. These net-like bryozoa form colonies large enough to be conspicuous fossils. This is particularly the case when as here, they are preserved in dark shales, the white calcite of the animal's skeleton standing out against the background. *Fenestrellina* species are wide-spread in Carboniferous rocks, and colonies are generally 5–6 cm across, but may be 10 cm or more.

14 Trigoniid bivalve, *Scabrotrigonia thoracica*, Cretaceous, Tennessee. A single valve of this species is exceptionally well-preserved showing the finest details of the shell structure. The radial ribs 'chopped up' into little knobs are characteristic of the trigoniid bivalves. The shell is very robust, and inside there are very few, powerful hinge teeth. The longest diameter of this specimen is $5\frac{1}{2}$ cm.

15 Carboniferous bivalve, *Aviculopecten planoradiatus*, Derbyshire, England. For a fossil of this age the preservation is remarkable, because the original colour banding is preserved, showing as broad, V-shaped patterns. The single valves are preserved in a fine-grained lime-stone. Pectinid shells have grooves widening towards the margin, and, at the apex, flattened 'ears'. Length, 3 cm.

16 The gastropod *Desmoulia conglobata*, from the Pliocene of Italy. As befits its relatively young geological age, the specimen is preserved with its original shell material. It has been sliced longways to show the internal structure. Note the *columella*, which is twisted like a corkscrew. Length, 4 cm.

17 Jurassic nautiloid, *Cenoceras pseudolineatus*, Dorset, England. Specimen largely preserved as an internal mould. A section cut and polished through the specimen (on the right) shows the internal chambers filled or partly filled with calcite. The internal mould shows the gently-curving suture lines. The specimen has a diameter of 7 cm.

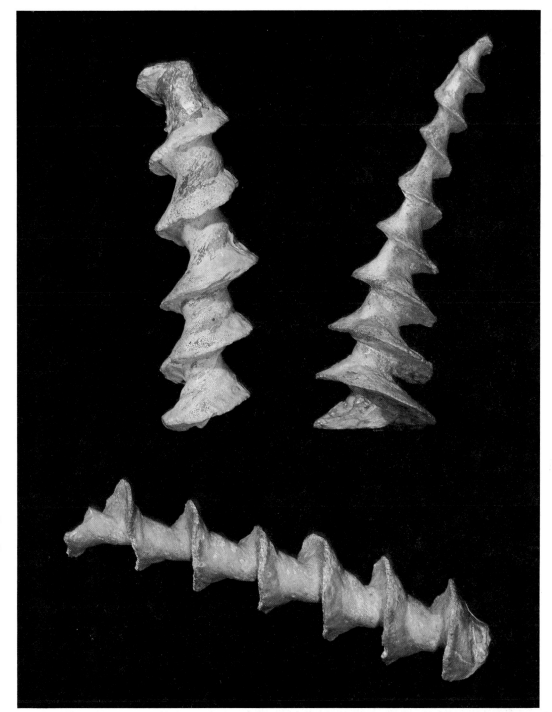

Peculiar, screw-like bryozoan, *Archimedes sublaxus*, Carboniferous. The calcareous skeletons of this distinctive species have weathered out from the limestone matrix. While most bryozoa require microscopic examination, a few form colonies large and distinctive enough to be easily recognizable. This is one such form, with its colonies forming peculiar miniature helter-skelters. The genus is named after the famous Greek philosopher, and inventor of the 'Archimedes screw'. Numerous *Archimedes* species occur worldwide in rocks of Carboniferous and Permian age. The colonies grow to a length of 5 cm. Larger species have been described.

over long distances, and are quick to colonize vacant sites that appear in the ocean (new volcanic islands like Surtsey, for example). Most molluscs are small, a few centimetres long, and some are really tiny, but a few species have attained considerable dimensions. The most famous (or notorious) is the giant squid at 15 m or more in length, but some of the extinct ammonites and nautiloids were of similar dimensions, and were the largest shelled animals ever to have lived.

Because the molluscs are such a large and varied group, with such an extensive fossil record, we shall consider them below in their various classes. Accurate identification of molluscs is a skilled business, and there is a vast literature describing fossil species, but it is easy to recognize the most important types preserved as fossils. The fossil record, particularly in the Cambrian, is still turning up new and fascinating kinds of molluscs, some of them really weird, and a few of these will be touched upon after the most important and familiar groups have been described.

THE CLAMS – CLASS BIVALVIA (Colour plates 14, 15)

In clams the body is enclosed in a pair of valves, which in most species are mirror images of one another. The valves are composed of calcium carbonate, are quite strong in most species, and hence easily fossilized. Between the valves a springy ligament keeps the valves in a gaping attitude – the position used for feeding. If danger threatens powerful muscles can snap the valves tightly shut, and once a bivalve has closed itself in this way it can be very difficult to force it open – it has 'clammed up'. The strong muscles leave scars on the interior of the shell at the places to which they were attached (usually two in each valve), and these muscle scars may also be seen on fossil shells. To make the hinge strong and efficient there are usually complex arrangements of teeth and sockets there, and the arrangement of these articulating devices is very important in identifying the different kinds of clams. So it is important to find the internal structures in fossil bivalves, as in the brachiopods. The clams use their foot for digging and movement generally. The group has adapted to a range of marine and fresh water habitats where they are often filter-feeders. Their fossils often occur gregariously as they lived, forming beds largely composed of fossil shells. Many living species burrow into sand or mud, sometimes to a considerable depth – these species maintain contact with the sea by means of long *siphons*, tubes that permit the passage in and out of water and bring to the animal both the necessary oxygen and the small organic particles on which it feeds. Other species are attached to rocks by means of tough threads (*byssus*) that enable them to hang on even in turbulent situations; mussels can colonize the most inhospitable rock surfaces in this way. A few bivalves have become free-swimming – such pectinids can escape predators rapidly by 'clapping' their valves together. Still others have taken to burrowing into wood or even into limestone, and fossils of these curious animals can be found lying in their home-made burrows (*see* p. 21). With such a wide range of adaptations it is not surprising to find that the shapes of bivalves are highly varied – some are globular, others flat and plate-like, some like the razor shell (*Ensis*) have become greatly elongated to aid burrowing, and in some forms the usual 'mirror image' symmetry has been lost. The thicker shelled

Tellin bivalve, *Tellinella rostralis*. Eocene. The original shell is here preserved in a sandstone. In life the tellin would have had two valves, but only one is preserved here. This species has an elongate form, which is extended at one end into a shovel-like tongue; fine concentric ribs are nearly parallel to its margin. Species somewhat resembling that shown are numerous in the Tertiary marine formations, and similar species live today in sandy sea bottoms. This species is from the Eocene of Belgium; related species are very widespread. The specimen is about 3 cm long, but may grow larger species often carry a distinctive sculpture, which is also important in identifying fossil species.

Most bivalve fossils are a few centimetres long; the ideal size for collecting. But a few giants are known. The Cretaceous genus *Inoceramus* sometimes grew to well over a metre in length; fragments of this particular genus are frequently important components of the soft, white Cretaceous limestone known as chalk. The giant clam (*Tridacna*) of modern reefs is a familiar living goliath, but reports of it greedily trapping divers are legends of dubious veracity.

The bivalves have a long geological history, with a few doubtful species known even in Cambrian strata, but like many other molluscan groups they really become established and diverse during the Ordovician. By the end of the Ordovician they had already radiated into many of the niches they occupy today. Their story from then on is one of slow, but steady, increase and diversification. Bivalves seem generally to have evolved at a slow canter rather than a brisk gallop (*see* ammonites pp. 73–6) and some living bivalves have a very long ancestry; the small-sized genus *Nucula* has relatives in Ordovician rocks not very different from Recent species. But the ancestors of many of the modern bivalves arose during the Mesozoic, and unlike many of the animals in this book the bivalves were affected in a much

less dramatic way by the Mesozoic–Tertiary extinction events that extirpated so many other major elements in the marine fauna. In spite of their general conservatism the bivalves *did* produce some short-lived, bizarre forms with no living survivors. Most extraordinary of these are the Cretaceous rudists (p. 47), a group in which one valve became modified to a long cone, on which the other valve rested like a lid, the whole effect being most un-clammish.

Some clams have shells composed of the form of calcium carbonate known as aragonite (like scleractinian corals) which is easily dissolved away when they are entombed in sediment: this leaves the clam preserved as an impression on the sediment of its internal and external surfaces (internal and external moulds). It is important to collect both 'halves' of the fossil to get an accurate representation of the shell – the external mould will preserve the overall shape and sculptural details, while the internal mould shows what the teeth and muscle scars were like.

The reader may be confused to find what we here call bivalves, referred to in other books as pelecypods or lamellibranchs – these are just different names for the same animals. The name accepted by most scientists today for the two-shelled class of molluscs is also fortunately the simplest – Bivalvia.

THE SNAILS – CLASS GASTROPODA (Colour plates 16, 18)

'Slugs and snails and puppy dogs' tails' – there seems to be something in the popular imagination that finds snails slightly repulsive. But this class of molluscs includes not only the greatest number of living molluscan species, including those that have most successfully colonized land, but also some of the most beautiful examples of natural engineering in the zoological world. Some of their shells have a financial value that may even be out of proportion to their aesthetic qualities. The gastropods are molluscs with a single, usually helically-coiled shell, with the foot modified into an efficient creeping organ, with a head, usually with eyes and tentacles, and with a rasp-like feeding organ (*radula*) composed of a series of pointed teeth.

The gastropods have become adapted to a very wide range of habitats, from high mountain streams to deep oceans, and each habitat type has its own species confined to it. Fresh water gastropods are different from

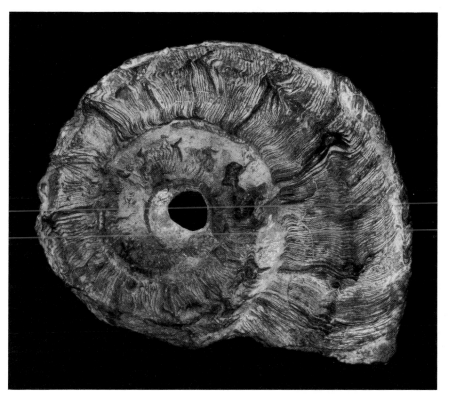

marine ones, and within the marine habitat itself the gastropods are strongly zoned ecologically, so that even on the same shore different species will be found in different areas according to their relation to the tide marks, degree of exposure, their diet, and so on. Usually such fine habitat details are not preserved in the fossil record, and the gastropods we find are a jumble of shells displaced from their original microhabitats. Many gastropods are grazers: they use their radula to rasp away at algae. Some feed more or less directly on sediment, from which they extract edible particles. There are a large number of living species that are predators, some species using the modified radula as a kind of poison dart, others employing it to bore neat, perfectly round holes through the shells of prey (often bivalves) in order to get at the nutritious interior. Some of these predators are specialists concentrating on one particular prey

Fossil limpet gastropod, *Symmetrocapulus tessonii*, Jurassic. The species is preserved with its original shell, its interior being filled with limestone. The fossil shows well the original concentric growth wrinkles. Fossil limpets of this kind are generally rare as fossils, and the amateur will be lucky to find one. This specimen is very rare indeed. It is from the Jurassic limestone of Les Moutiers en Cinglais, France. It is 11 cm long – large for a limpet-like mollusc.

Fossil sundial shell, *Architectonica millegranosa*, Pliocene. The well preserved shells of this gastropod have been washed almost free of their enclosing silty matrix. The low spire of this species, with a minutely beaded ornament, and the sharp rib around its outside edge, are features that discriminate this species from other gastropods. This species is one of a farily large genus that survives today. They are uncommon in the tropical or subtropical molluscan faunas of the Tertiary. Species resembling this one in general shape are met with in rocks of Cretaceous and younger age. This species is from Orciano, Italy. Diameter about 3 cm.

Gastropod, *Harpagodes wrightii*, Jurassic. This large gastropod is preserved in an oolitic limestone. The long, stout spines are an unusual feature which discriminates this species from other gastropods. Note that the surface of this species bears encrusting oysters of a species that took advantage of the hard surface of the dead gastropod. This species is a real rarity. It was discovered in the Great Oolite limestones of Gloucestershire, England. A few related species occur in the Jurassic. The longest diameter of the specimen is nearly 15 cm. Cleaning out gastropods from oolitic lime-stones can be a lengthy business. Usually the weathering has done part of the job, but thereafter a combination of small chisels and stout needles (to prise away individual grains) is needed. Always remember to keep clear of the soft calcite shell with these tools.

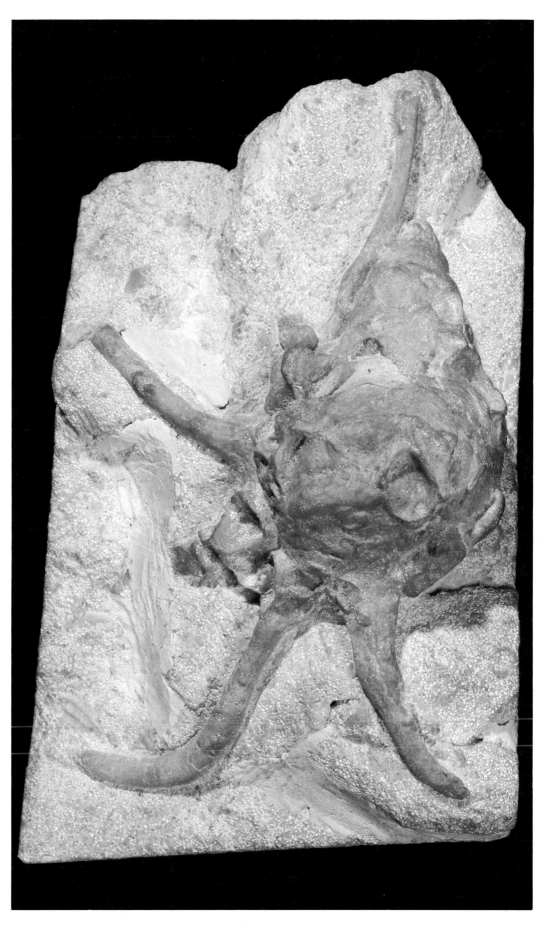

type. It will now be clear why you can sometimes find so many different fossil gastropod shells together in a single fossil deposit.

The gastropod shell can vary from a millimetre in length to several tens of centimetres; many are thick and robust, others delicate and fragile. While the majority are coiled in an upward spiral, some are modified into simple, cap-like forms (like the limpets), and others are coiled in a flat plane, like a ram's horn. Some genera have high spires, with many turns of the shell visible externally, others have low, broad spires in which the last *whorl* overlaps the earlier ones. Perhaps more than any other group of molluscs the gastropods are remarkable for

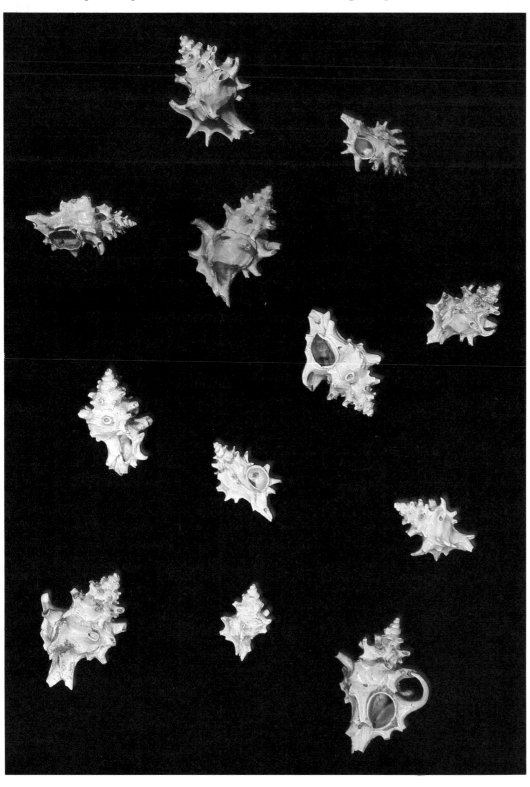

Typhis pungens, a fossil gastropod, Eocene. These specimens are well-preserved in a clay matrix, some of which adheres to one of the specimens. Such spiny little fossils are unlikely to be confused with other gastropods. Note also the oval aperture surrounded by a narrow rim. There are a number of *Typhis* species, with various numbers and arrangements of the spines. Fossil *Typhis* are frequent fossils in the Eocene rocks of Europe. *Typhis* still lives today – in the seas around Japan, for example. The series of fossils shown here demonstrate the different sizes gastropods of one species attain. The largest here is 3 cm long.

the variety and beauty of the external sculpture on the shell, which may be covered with a delicate tracery of ribs and lines, or stout spines, or fine prickles. Many species have the aperture flared, or extended into a long tube (*siphonate* forms). Sometimes you can find a lid (*operculum*) preserved as a fossil, that closed the aperture when the mollusc withdrew into its shell. All these characters are used in the identification of fossil species, but of course the colour patterns that may be characteristic of living species are not available in the great majority of fossils. Like some of the bivalves, some gastropods had shells composed of the mineral aragonite, and these are usually preserved as moulds: but a cast taken from the external mould will replicate the fine detail of the external sculpture.

There are small helically-coiled shells present even in Cambrian rocks, and it seems likely that the gastropods diverged from the other molluscs late in the Precambrian. By the Ordovician the gastropods were a varied group present in a variety of shallow water habitats. As might be expected most of the Palaeozoic gastropods belong to primitive groups: a few of these primitive snails survive today as inconspicuous members of the Recent fauna. By the Carboniferous many of the shapes we see in living gastropods can be matched in the

fossils, but despite these similarities the majority of the Palaeozoic forms were not closely related to their living analogues; this is another example of similar-looking forms evolving independently probably in response to similar life habits. It was during the Mesozoic that the forms ancestral to many of the living gastropods evolved. In particular the adoption of predatory habits was a new departure for the gastropods, and was in part responsible for a proliferation of the group unmatched in their Palaeozoic history. They continued to evolve with undiminished vigour through the Tertiary, and their fossil remains are nowhere more abundant than in the 'crags' of later Tertiary age. The gastropods are one of many groups that record the faunal changes connected with the advance and retreat of the Ice Sheets during the Pleistocene. The invasion of the non-marine habitat probably first happened in the Carboniferous, but relatives of the living land snails are rare before the Cretaceous, at which time the familiar *Helix* made its first appearance. The slugs were derived from this group by reduction of the shell, at which stage they become distinctly less qualified to have a fossil record!

Gastropods can leave other evidence of their activities. Grazing species leave characteristic winding trails, and these have been

The living pearly *Nautilus*; a section showing its internal division into chambers

tentatively identified as fossils, but with the usual caution that the trail makers themselves are never apparently preserved at the end of their tracks.

NAUTILOIDS, AMMONOIDS, BELEMNOIDS, SQUIDS – CLASS CEPHALOPODA (Colour plates 17, 19, 20, 21)

The cephalopods include the most complex and 'advanced' of the molluscs, and are a group of the greatest geological, as well as biological interest. They have muscular tentacles in a well-developed head region, highly efficient eyes which are similar in construction to those of vertebrates (although obviously independently derived), and they feed with the aid of strong, beak-like jaws. They are predators, and probably always have been, and it may have been the adoption of hunting habits that favoured the development of high intelligence. Certainly, living cephalopods have a sophisticated nervous system and a relatively large 'brain': octopuses seem to be capable of very rapid learning. Perhaps H. G. Wells made the right choice when he cast octopus-like animals in the role of intelligent alien invaders in *The War of the Worlds*.

The cephalopods evolved from another (gastropod-like?) mollusc at some time during the Cambrian, but their earliest history is little known. Like most other molluscan

groups they rapidly diversified in the Ordovician. These early nautiloids were probably predators also, and if this were so they may have been the first rapidly moving, efficient hunters in the sea. Their impact on the marine communities of the Ordovician must have been profound. The presence of efficient predators would have acted as a stimulus in the evolution of other groups: even algal grazers would be compelled to evolve protective devices or rapid reproductive strategies to outpace predatory depredations. We have seen how many animal groups diversified during the Ordovician; the rapid evolution of nautiloid cephalopods at the same time may be more than coincidence.

Rapid movement in cephalopods is achieved by expulsion of water from a muscular funnel beneath the head. Most of the living cephalopods also have an ink sac which injects a smoky fluid into the water when the animal is threatened, under cover of which they can make their own jet-propelled escape. Since living *Nautilus* does *not* have an ink sac, this must have been a protective device evolved at a later stage in cephalopod evolution. Many cephalopod species behave gregariously today, travelling in swarms in pursuit of their favourite prey (often small crustacea). Fossil cephalopods belonging to one species are often found together in large numbers and this may reflect similar gregarious habits, but there are other possible explanations – for example, concentrations of fossil shells may have been sorted by currents.

NAUTILOIDS

The earliest cephalopods found as fossils are nautiloids and they also have the longest history, because the living pearly *Nautilus* belongs to the same group. The single Recent genus scarcely reflects the diversity attained by the group in Palaeozoic seas, and more than a hundred different genera have been described from the Lower Palaeozoic alone. Sections through the living *Nautilus* (*see* p. 72) are often sold as ornaments, and well illustrate the distinctive features of nautiloid hard parts. The body of the animal occupied the cavity at the end of the spiral – the living chamber. Behind the living chamber the rest of the shell is divided into smaller chambers. As the animal grew it secreted a wall (*septum*) between the body chamber and the chamber immediately behind – walled off part of its home as it were. Running along the middle part of the shell there is a narrow tube – the *siphuncle* – which connects the living chamber with the earlier parts of the shell. The empty chambers are usually supposed to have been filled with gas, which help to give the animal bouyancy, and via the siphuncle the animal can vary its bouyancy to control its position in the water column. Some of the early nautiloids deposit calcium carbonate in the voided chambers, which may also have been connected with controlling buoyancy. Where the septum meets the body wall it does so in a smooth curve. Internal moulds of fossil nautiloids often reveal a series of such lines marking the boundaries of the chambers – these are known as *suture lines*. All nautiloids have simple suture lines.

When they first appear in the early Ordovician the majority of nautiloid shells are straight or slightly curved; they are 'unwound' forms. Some of these straight orthocone nautiloids achieve considerable dimensions, several metres long, and they must have been formidable predators on the other marine animals of the time. Quite early in their history some partly coiled or even tightly coiled species evolved, and the various coiling types seem to have coexisted successfully side by side. Some of these early nautiloids occurred in such abundance that they are conspicuous enough to form an appreciable part of limestone formations –

the 'Orthoceras Limestone' (Ordovician) is one of these, widely distributed through Scandinavia. The nautiloids achieved their widest range of adaptations and greatest variety of form in the Ordovician and Silurian periods, with various coiled forms, straight, pipe-like forms, and curious dumpy species with restricted apertures that may have adopted a sluggish (possible filter-feeding) mode of life. Thin sections show a great variety of internal structures important in accurate identification. In the Devonian period the nautiloids were still abundant and varied, but they suffered a slow eclipse coincident with the rise of the ammonoids. Nonetheless unlike many of their Palaeozoic companions, they survived the late Permian extinction, and the ancestors of the living *Nautilus* even underwent a minor evolutionary burst in the Mesozoic, where forms quite similar to the pearly *Nautilus* can be common fossils. And in the end the nautiloids even survived the ammonites, the molluscan group that evolved more rapidly and more spectacularly than any other.

AMMONOIDS ('AMMONITES')

The ammonoids were derived from the nautiloids probably during the early Devonian, and from the Carboniferous until the Cretaceous are among the most abundant of fossil groups, in some rock types dominating to the exclusion of most other members of the fauna. Ammonoids differ from nautiloids in the suture lines being wavy or crimped; this of course reflects an elaboration of the outer part of the walls (septa) separating the chambers in the earlier part of the shell. The siphuncle in most ammonoids runs not through the middle of the whorl but along the outer edge. There has been a lot of debate about the reasons for producing complicated patterns on the suture line; what advantage would this have given the ammonoids over their nautiloid ancestors, that allowed for the explosive bursts of ammonoid evolution? It cannot be coincidence that the folding of the septal walls occurs at the point where they meet the body shell of the animal; this is a point of relative weakness, and all good joinery benefits from strengthening the joint. The earlier ammonoids had gently wavy sutures, and many of the Mesozoic ones had sutures almost incredibly folded and contorted, so it looks as if natural selection were generally favouring increase in the elaboration of the folds. One simple explanation, which is attractive

Ceratite ammonoid, *Ceratites nodosus*, Triassic. This specimen is an internal mould in limestone, and shows the suture lines well – the boundaries between chambers on the inside of the shell. This ammonite has only a few strong ribs, with a number of small tubercles in addition. The looping form of the suture lines is also characteristic. Ceratites are typical of Triassic rocks and there are numerous related species differing in details of ribbing and suture lines. Such ammonites are plentiful in marine limestones of the right age across the continent of Europe (though not in England). Related forms occur in America. The specimen shown here is from Göttingen. This specimen has a diameter of 6 cm but some are larger.

Jurassic ammonite, *Promicroceras planicosta*. Numerous small specimens are preserved higgledy-piggledy in an impure limestone. Some retain their shell, others show the dark internal moulds. Small size, unbranched and strong ribs, giving the shell a ramshorn appearance, are important characters. This form has a number of related species, but the situation is complicated by the fact that some large ammonites have inner whorls looking rather like *Promicroceras*. This specimen is from the Lias of Somerset, England. Individuals do not exceed 2 cm in diameter.

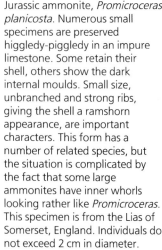

and plausible, is that the strengthening enabled the ammonoid shell to withstand the hydrostatic pressure at relatively great depths in the ocean – they need not be confined to the surface waters or to relatively shallow depths. Certainly individual ammonoid species became extremely widespread, and oceans were not a barrier to their distribution. They became masters of the pelagic realm, possibly swimming in schools like their distant relatives the squids. Like living pelagic animals they had preferences for particular water temperatures; different types of ammonoids were found in high or low latitudes. Since the disposition of continents has changed, these 'faunal realms' offer a method of deducing the distribution of past climatic belts.

Whatever the reasons for their change from their nautiloid ancestors, the ammonoids were an enormously successful group: thousands of different species have been described, and their variety is so bewildering that many specialists devote their lives to studying only the ammonoids of a particular, short time period. The changes that they underwent are an infinite set of variations on a relatively limited number of themes. Most important, perhaps, are the characteristics of the suture lines, which are displayed on internal moulds. The mature shell size also varies from small species a few centimetres across to giants of a metre or more in diameter. Some species grew in such a way that the last whorl overlaps the inner ones; often these species become flat and discus-shaped: it has been suggested that

this was a more 'streamlined' shape for active swimmers. Other species have squat whorls, the whole ammonoid being so tightly rolled up as to be almost spherical. The exterior surface of most ammonoids is covered with ribbing – dense on some species, sparse on others – the ribs often split into two or more smaller ribs as they pass over the back of the whorls. Many ammonoids additionally carry spines, warts, tubercles or lumpy excrescences, so that the large shells can look positively burdened with sculpture. It is difficult to imagine that such species were rapid swimmers. Others, particularly the discus-shaped forms, were smooth externally. All these features are used in the classification of the group. A particularly puzzling aspect of the ammonoids is a great resemblance between external shell features that can be produced at different times by otherwise unrelated ammonoids. Such *homoeomorphs* are presumably produced in response to very similar life habits. The origins of a particular homoeomorph can usually be deduced by studying the internal features (in particular the suture lines), or by tracing its derivation from geologically older species in underlying beds.

It is only in relatively recent years that two sexes have been recognized in ammonoid species. They were originally thought to have been two closely related species occurring together in the same rocks, but these 'pairs' were so consistently found together that it became more and more probable that they were sexual forms of the same species. The smaller of the two is considered to be the male, which also tends to have rather stronger ribbing, and sometimes a differently shaped aperture from the larger female.

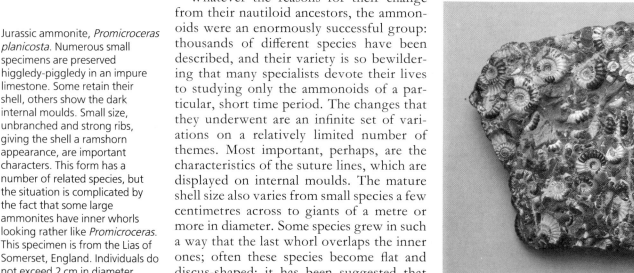

A group of belemnites, *Acroceolites subtenuis*, Jurassic. This fine group of belemnites is preserved on a soft shale. The belemnite guards themselves are composed of calcite. There are very many belemnite species in Jurassic and Cretaceous rocks, sometimes occurring in vast numbers. They require expert determination. The specimen shown here is from the Jurassic shales of Yorkshire, England, but similar species occur worldwide. The longest specimen shown here is 9 cm long. Some belemnites are as much as twice that length.

The early ammonoids of the Devonian (*Anarcestes* and allied genera) have only gently sinuous suture lines. In the Carboniferous a great variety of forms with highly zig-zag suture lines are a distinctive group of ammonoids usually known as *goniatites*. By the Permian–Triassic the suture lines of many ammonites had begun to assume the highly crimped and complex form that was to characterize much of their later history; the ceratites of the Triassic combined broad loops and tight folds in a distinctive pattern (*see* p. 74). It was in the Jurassic that the ammonoids achieved their greatest flowering, when the clays and shales of the period may be solid with the remains of their shells. The whole gamut of shell shapes, ornaments, and sutural complexity was present during this time. Their rise was not, however, one of simple and progressive increase in variety. At several times, notably at the end of the Palaeozoic, they suffered massive and largely unexplained extinctions, a few survivors giving rise to the variety of forms that followed. In spite of their success they seemed to be vulnerable to extirpation in a way that some of the less imaginative molluscs, like the bivalves, seemed to be immune. The group persisted, successfully, into the Cretaceous, at the end of which period the whole group, apparently rather suddenly, became extinct. We shall return to this sudden death of an important group in Chapter 6.

No description of the ammonoids is complete without mentioning the *heteromorphs*. These are forms which abandoned the usual plane spiral mode of coiling, and instead became partially or even completely *un*coiled, or became twisted in some other fashion. Some forms (*Turrilites*) adopted the helical spire, and were it not for the obligatory suture lines it might be possible to mistake these species for large gastropods. In others the coils became loose, like a watch spring. Still others developed a bizarre backward hook in their mature stage. The extreme heteromorphs are perhaps to be found in the genus *Nipponites*, which looks like a tangle of whorls where any obvious semblance of symmetrical coiling seems to have been lost (*see* left), and *Baculites*, which is virtually straight after its earliest whorls. Most of these heteromorphs were derived from 'normal' ammonoids, but there is one

Anomalous ammonites: the Cretaceous genus *Nipponites*. About × 1½.

famous example where an uncoiled form actually gave rise to a conventional-looking ammonoid by coiling up again! Heteromorphs are particularly common in, but by no means confined to, Cretaceous rocks. At one time it was supposed that the ammonoids were suffering from 'racial senescence' at that time and that the uncoiling represented a kind of genetic exhaustion. This resulted in their reversion to resemble some of their earliest ancestors (the straight nautiloids) and squared with the observation that some of the heteromorphs even reverted to simple, wavy sutures again. However, the heteromorphs were extremely successful and widespread, and they are acccompanied by, and even survived by other species with a perfectly usual appearance. The explanation lies in something less mysterious than 'racial senescence'. Far from being an expression of decline the heteromorphs show the ammonoids adopting new (and successful) life habits. Some of them may have become bottom living, crawling hunters for which a gastropod-like shell would have been more appropriate. Loss of active swimming habits may have rendered the complex suture lines superfluous. But their displacement from a dominant role in the pelagic habitat may have accompanied the rise of squid, sepioids, and octopods.

Ammonoids, heteromorphs included, are almost without parallel as stratigraphic indicators. They evolved rapidly and spread widely, and have a range of distinctive characters to help the investigator in his identifications. Study of evolving populations of ammonoids has produced very fine subdivisions of Jurassic and Cretaceous rocks, and their only disadvantage, curiously enough, is the feature that makes them so attractive to collectors. They are large, and hence the chances of recovering complete specimens from all boreholes is relatively low. Microfossils (Chapter 8) are often used in their stead for subsurface work.

BELEMNOIDS

The cigar-shaped belemnoids ('belemnites') are common fossils accompanying the ammonoids in Jurassic and Cretaceous rocks. The fossil is an internal skeleton (enclosed within the body) of a squid-like animal, in a way comparable with the cuttle bone of the living cuttle-fish. The solid calcite of this internal skeleton makes the belemnoid a resistant fossil, and fragments are common survivors of erosion, often picked up on 'Recent' beaches. The chambered shell (*phragmocone*) is less conspicuous – it is tucked into the broad end of the fossil guard, where the series of closely-spaced septa reveal the cephalopod nature of these otherwise somewhat featureless fossils. Belemnites vary from small fossils a centimetre or two long to large specimens tens of centimetres long: of course these are only a fraction of the size of the living animal, with their tentacles extending well beyond the guard. The belemnoids disappear from the fossil record at the end of the Cretaceous, but some of the group gave rise to living squid-like animals, and so they are not to be regarded as extinct in the same, final way as the ammonoids.

OTHER MOLLUSCAN CLASSES

Some molluscs are found as fossils which belong to groups other than the familiar ones described previously. Although they are rarer fossils, some of them are of such interest that they are worth mentioning here. One of the most exciting are the Monoplacophora. These are a group of extremely primitive molluscs, which are found in any abundance only in Lower Palaeozoic rocks. They are usually simple cap-shaped shells looking somewhat like limpets, but on their internal surfaces they carry a series of paired muscle impressions. Many authorities regard the Monoplacophora as lying at the root of the other molluscan groups; gastropods, cephalopods and even bivalves may have been derived from them. For many years they were known only from fossil representatives and were believed to be extinct. It was an amazing discovery to find that monoplacophorans were still alive and well – living ones were dredged from the deep ocean in the early 1950s. Not only that, the living form had hardly changed from its Silurian predecessor (*see* p. 77). The monoplacophorans really reached their acme in the Cambrian, with curious forms having shapes that have never been paralleled in other molluscs (*Yochelcionella*).

Another odd group that did not survive the Palaeozoic were the Rostroconchs. These look rather like bivalves without a hinge. They really are extinct, but managed to compete successfully with the bivalves for a considerable time.

Finally, the tusk shells (Scaphopoda) are a long-lived group with a record stretching from the Ordovician to the present. They are the only molluscs with a truly tubular shell, usually gently curved, with an open-

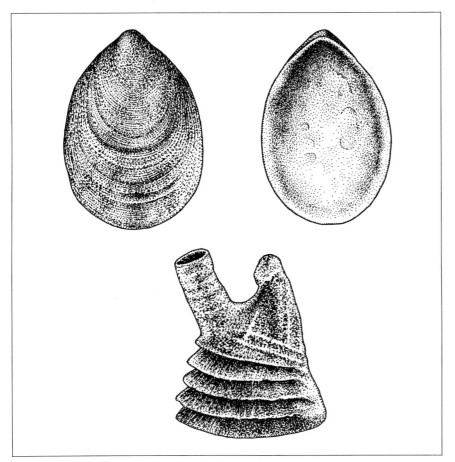

Primitive molluscs: Mono-placophora. Silurian fossils very like the 'living fossil' *Neopilina* (*above*), and the peculiar Cambrian form *Yochelcionella* (*below*).

a relatively strong assembly of calcite plates; the animals are built from an interlocking mosaic of such plates, and mostly the skeletons are rigid enough to have a high chance of fossilization. The geological record of the phylum is accordingly excellent. The echinoderm skeleton is not a truly external one, like that of the molluscs, for it is surrounded on the *outside* by a thin skin of living tissue. The mosaic of calcite plates is a particular feature of the phylum, and serves as one of the characters linking together such dissimilar-looking animals as starfish and sea lilies. Another unique feature is the water vascular system, a system of internal plumbing that drives the *tube feet*: mobile, club shaped sacs arranged in serried ranks on the outside of the animal, which operate in harmony to convey food to the mouth, or in locomotion. They have a well-developed nervous system. Most echinoderms also have an unique five-rayed radial symmetry in the overall shape of the skeleton. This five-fold (pentameral) symmetry seems a peculiar number to be present over such a wide range of organisms. Why not four, six or thirteen? The answer is still not clear, but what is certain is that the five-rayed plan was established even in the Cambrian in a number of diverse echinoderms; it is evidently highly functional.

In detail the echinoderms show evidence of bilateral (mirror image) symmetry, from which five-fold symmetry was probably derived. Once acquired it is rarely lost, even in echinoderms like sea urchins that (perhaps) could also function on a different symmetry pattern. Each plate of an echinoderm is composed of a single crystal of calcite, whereas all of the organisms we have discussed above have skeletons composed of felt-like masses of tiny crystals. Broken echinoderm plates 'catch the light' – the single crystals break along cleavage planes and so present a uniform reflective surface. Many rocks are composed of a high proportion of echinoderm debris, which is easily recognized because of this optical property. Through their long history the echinoderms have never left the sea, but within their preferred medium they have adopted most of the life habits available to marine organisms. They include grazers, feeding on simple plants, scavengers, mud eaters, filter-feeders extracting micro-organisms from currents, and some efficient and versatile hunters. Like other marine organisms they have a planktonic larval stage that assists their dispersal, but almost all adult echinoderms lead a bottom dwel-

ing at both ends. They live today with the broad end buried in the sediment, where they forage for food using small prehensile filaments. Their habits have probably always been similar, and if survival is to be taken as a measure of success, their conservative way of life has ensured them of a leading place in the evolutionary marathon. Other animals described in this book have occupied a specific ecological niche where they can quietly pursue their own speciality without embarking on any spectacular radiation in the manner of the ammonoids. It is a curious paradox that often the most spectacularly successful and numerous organisms are also those with a finite geological record.

THE ECHINODERMS – PHYLUM ECHINODERMATA

Sea urchins, starfish, sea lilies and sea cucumbers are living echinoderms – the 'spiny skinned' animals, according to the Greek name of the phylum. It is an appropriate name, for most echinoderms do feel prickly to the touch, and the sea urchins are equipped with fearsome spines. The skeleton of all echinoderms except the sea cucumbers is ling existence; a few sea lilies have successfully cast off their anchorage to the sea floor and become widespread swimmers.

The echinoderms are such a varied and important group that the different kinds have to be considered separately. I should add that some of the most interesting echinoderms are also some of the rarest – peculiar, plated animals that do not fit comfortably into the array of living forms. Many of these occur in the Cambrian – it is as if the echinoderms tried out various designs before settling for the successful models that mostly survive today. The echinoderms are a phylum closely allied to the Chordates – including the vertebrates, and both phyla probably have a common ancestor in the Precambrian.

SEA LILIES – CLASS CRINOIDEA, AND OTHER STEMMED ECHINODERMS (Colour plates 22, 23, 24)

Crinoids are abundant and important fossils from the Ordovician to the Tertiary. They are still abundant at present, although a little less common in shallow water sites than they were in the Palaeozoic and Mesozoic, but they are conspicuous and varied components of deep water faunas. The great majority of the group have long stalks, which are anchored to the bottom. The main part of the animal consists of a cup (or *calyx*) to which the stalk is attached at its upper end, and from the top of the calyx stretch long arms, which are five in number or more usually a multiple of five. The arms are often repeatedly branched. The distal parts of the arms carry fine *pinnules*, which are instrumental in gathering the fine, water-borne food such as micro-organisms on which the animal feeds. Since crinoids often occur together in large numbers ('gardens') with their arms waving in the currents, it is easy to see how they came to acquire their botanical analogy. Food grooves in the arms channel the food to the mouth, which lies in the centre of the calyx. Stem, calyx, and arms are all made of calcite plates. The stems are often common fossils, even when the calyx cannot be found – many of the 'crinoidal limestones' are composed largely of stem debris. The individual plates of the stem slot together like a stack of coins; they are termed *ossicles*. These can be round, or five-sided, up to a centimetre or so across, and have a small hole in their centre; the stems can look like corals to the casual observer, but where they are broken they show the typical echinoderm reflective

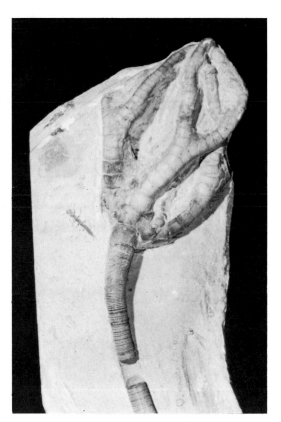

Crinoid, *Onychocrinus exsculptus* from the Carboniferous (Mississipian). The species is preserved in relief in limestone, which is very fine grained, and retains many fine details. The long stem of this crinoid, with a small cup at the top, and stout arms composed of easily visible calcite plates, these first arms themselves dividing, are some of the more important characteristics of this species. There are numerous Carboniferous crinoids not unlike this one, which require expert knowledge in their discrimination. This specimen is from Illinois, USA, but forms related to it may be found in Carboniferous limestones in many areas. The specimen is 15 cm long.

surfaces. Detailed classification of the crinoids is based principally on the way the calcite plates are arranged on the calyx with the number and branching patterns of the arms, the external sculpture on the calyx, and important internal features of the cup connected with the nervous and respiratory systems. In detail they are a very complicated group.

Complete specimens of fossil crinoids are rather rare, and deserve pride of place in anyone's collection. Complete cups are more common – many of them were evidently rigid, and they are easily preserved as fossils. There are famous examples where whole 'gardens' seem to have been preserved, with the arms frozen, as it were, in motion, and even the roots at the base of the stems in place.

The crinoids have had an eventful geological history. True crinoids are doubtfully known before the Ordovician, but once established they diversified rapidly in the manner we have seen repeatedly with other groups. They soon spread to a variety of habitats, but in the Palaeozoic they were conspicuously abundant in relatively shallow environments. The most mouthwatering specimens tend to be found in limestones – famous sites are in the Silurian rocks of England and Mississipian rocks of the U.S.A. The group as a whole had a

Fossil crinoid (sea lily), *Dizygocrinus montgomeryensis*, Carboniferous (Mississipian). The specimen is unusually complete for a fossil sea lily, and lies on a fine-grained limestone. There is a certain amount of crushing of the cup beneath the arms. The stem is broken off. Note the fine, thread-like *pinnules* attached to the arms – used by the animal to trap food particles – and the thin and flexible stem. There are quite a large number of species in Carboniferous and Permian rocks which resemble the species illustrated here. This species is from the limestones of Keokuk, Iowa, USA. The arms of this specimen have a length of about 4 cm.

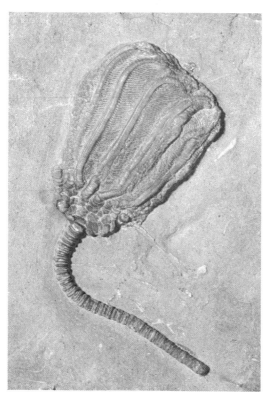

Blastoid, *Pentremites spicatus*, Carboniferous (Mississipian). The specimen is preserved in its original calcite; its lack of distortion may be seen from its perfect five-fold symmetry. The five finely striated food grooves, which in this view occupy much of the surface of the animal, are typical of the blastoid. The name 'blastoid' is derived from the Greek for 'bud' because of the resemblance of the animal to a flower bud. *Pentremites* is a large genus with many species in limestone of Carboniferous age, particularly in the Mississippian limestones of North America. This specimen is from Grayson Co., Kentucky, USA. Similar species occur in North and South America, and related genera almost worldwide. The specimen has a diameter of 3 cm.

major crisis in the Permian, during which most of the Palaeozoic forms died out. A very few of these survived into the Triassic. In the Mesozoic there was another great radiation of the crinoids, the typical forms having flexible arms, and it is crinoids of this kind that survive today. Some of the most famous crinoidal deposits are of Jurassic age, and marvellous specimens have been collected from the Lias rocks of Britain and Germany. A new and successful innovation was the evolution of stemless crinoids – some of these acquired pelagic habits, and during the Cretaceous the genera *Marsupites* and *Uintacrinus* (*see* p. 79) were widespread enough to be useful marker fossils. Some of these stemless forms developed grappling hooks at the base of the calyx – with these they can attach in a favourable site, and move when conditions become adverse. Such feather stars are very numerous in some reef habitats today. The long and varied history of the crinoids demonstrates how well the echinoderms have attacked the problems of filter-feeding.

In the Palaeozoic rocks there were other stalked filter-feeding echinoderms, which failed to survive the crisis at the end of the Permian, weathered by the crinoids. These extinct groups can exceed crinoids in number and variety at certain horizons, and they were evidently competing on equal terms. The *blastoids* (class Blastoidea) had

compact cups up to a few centimetres long, with five broad food grooves running down the *sides* (lacking crinoid arms). In life the food grooves were flanked by lines of delicate armlets (*brachioles*) which served to gather the food. Blastoids are sometimes abundant fossils in rocks (usually limestone) of Silurian to early Permian age, and their perfect, compact pentameral symmetry makes their calices among the most attractive of fossils. Like crinoids they evidently grew in gardens, for their remains (especially in Carboniferous rocks) tend to occur packed together in thousands. *Cystoids* (class Cystoidea) are even odder animals: often rather irregular bags of calcite plates, or if composed of a few plates these may carry powerful ribbing. The five-fold symmetry is often hard to detect in the calyx as a whole, although five food grooves are usually developed. They are a primitive, but very interesting, group with Cambrian origins, and reached their greatest variety in the Ordovician. Some of them evidently lacked a stem, and must have lain loosely on the bottom. There are several more groups of odd echinoderms in the Lower Palaeozoic rocks – some of them only recently discovered, like the bizarre helicoplacoids, which look like nothing so much as spinning tops (*see* p. 84). All are of the greatest interest, and *any* plated animal from the Cambrian is likely to be an important specimen. Sometimes these early forms are found as moulds – the original calcite having dissolved away. The external moulds then present a characteristic appearance, like the impression of a mosaic pressed into the rock. It is well worth keeping a

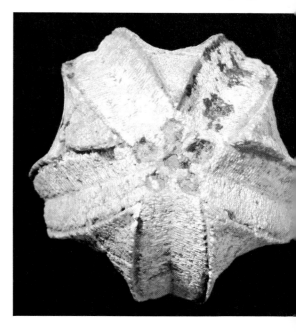

18 Eocene gastropod, *Voluta muricina*. The species is beautifully preserved, retaining something of its original lustre, and all the fine details of its ornament. It is distinguished by its tall spire, elongate aperture, prominent spines, but without the spiral ridges seen on many species. The specimen here is from Epernay in France, and related forms can be found in Eocene clays over much of Europe. Length, 7 cm.

19 Carboniferous goniatite, *Goniatites crenistria*, Derbyshire, England. This handsome goniatite is an internal mould preserved in limestone. The mould reveals the *suture lines* which in goniatites have a distinctive zigzag form. The aperture is broken off along one such suture, giving the specimen a pointed margin. The last whorl conceals most of the earlier ones. Diameter 5 cm.

20 Ammonite, *Kosmoceras acutistriatum*, Jurassic, Wiltshire, England. This ammonite occurs in a fine-grained grey shale, but has been severely flattened. The original lustre of the shell has been retained. This species is particularly distinctive because of the extended flanges at its apertural margin (*lappets*). Longest diameter of this specimen is 9 cm, but some forms grow larger.

20   **18**   **19**

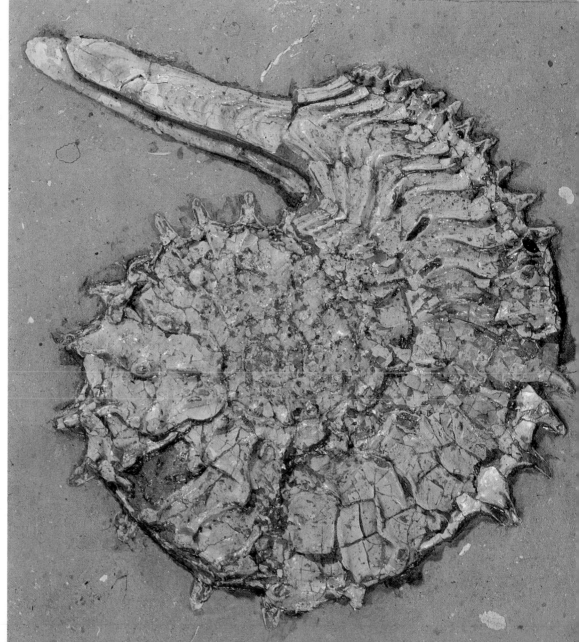

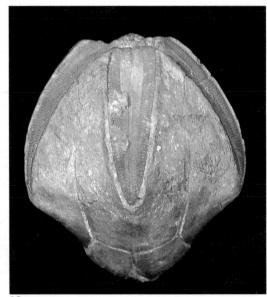

21 The heteromorph ammonite, *Scaphites nodosus*, Cretaceous, Badlands, South Dakota, USA. This species retains much of its original shell, giving it a beautiful pearly lustre. This is one of the later ammonites, in which the normal ammonite plane spiral has begun to 'unwind' in various ways. This species has a distinctive combination of fine ribbing and coarse tubercles. Long diameter of this specimen 8 cm; smaller examples are perhaps commoner.

21   **22**

22 Blastoid, *Pentremites robustus*, Carboniferous (Mississipian), Illinois, USA. The specimen is preserved in its original calcite with only the slightest compression. The view shown is from the side. The length of the illustrated specimen is 4 cm. In some localities blastoids like this one break easily out of the enclosing matrix, or are weathered out in considerable numbers.

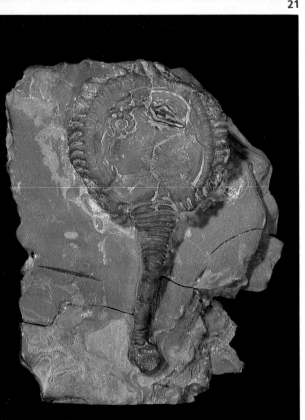

23 The Silurian cystoid, *Pseudocrinites magnificus*, Wenlock limestone, England. The calcite skeleton of this species is preserved in a grey limestone-shale matrix. The individual calcite plates of which the fossil is composed can be clearly seen. The stem is composed of a number of rings. This rare and peculiar fossil looks rather like a sea-lily without arms, and has more than a passing resemblance to a tennis racket! It has a length of 5 cm.

24 Triassic crinoid, *Encrinus liliiformis*. The calcite of this species is preserved with a gloss on the surface, in a matrix of slightly cavernous limestone, which is composed of many broken fragments of other fossils. *Encrinus* is typical of the marine limestones of the Triassic of Europe. This specimen is from the Muschelkalk of Germany. Total length 18 cm.

23

24

25 Cidaroid sea urchin, *Plegiocidaris coronata*, Jurassic, Ulm, Germany. Two beautiful examples of this sea urchin are illustrated preserved in a fine-grained limestone in full relief. Note however, that the stout spines are not preserved on these specimens. *Plegiocidaris* is found in Triassic and Jurassic rocks of Europe. The specimens have 5 cm diameter.

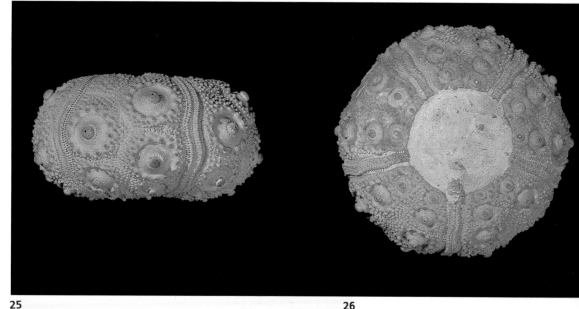

25   **26**

26 Jurassic sea urchin, *Hemicidaris intermedia*, Wiltshire, England. This specimen is exceptionally well-preserved, retaining its spines where they fell around the animal as it died. The matrix is a yellow limestone. The long spines of this species are characteristic – they resemble lances used in medieval jousting. Note the prominent articulating tubercles, like those of other cidaroids. The longest spine on this specimen has a length of 7 cm.

28 Spatangoid sea urchin, *Micraster coranguinum*, Cretaceous. This specimen is preserved in its original calcite. It is frequently found also as siliceous internal moulds. *Micraster* is sometimes called the heart urchin, because of its distinctive shape. Most specimens have a length of 5–7 cm.

27 **28**

27 Sea urchin, *Tylocidaris clavigera*, Cretaceous, Gravesend, England. Magnificently preserved in its original calcite in a matrix of chalk, this specimen is unusual because it has the spines still joined to the rest of the animal. The ferocious-looking spines, which seem almost too large for the central animal, are mostly attached, but one or two have come adrift. 8 cm across.

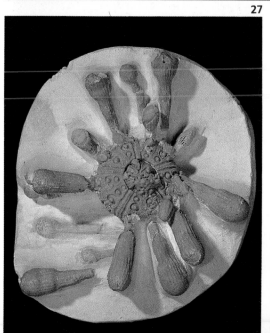

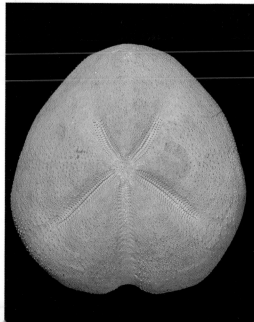

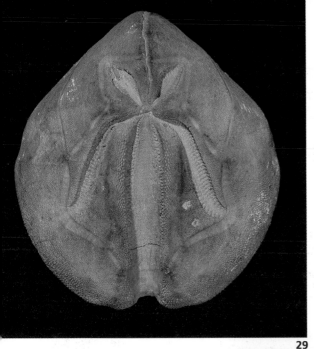

29 Sea urchin, *Schizaster canatifera*, Pliocene, Perpignan, France. The specimen is perfectly preserved – retaining even the finest details of its external surface. But the fine fuzz of tiny spines that covered it in life are not present. This species looks rather like the heart urchin but the petal-like areas are much more deeply sunken, the forward pair very short. Related species are worldwide. It is 8 cm long.

30 Fossil starfish, *Archastropecten cotteswoldiae*, Jurassic, Gloucestershire, England. The original calcite of the starfish is stained slightly yellow on the surface. The specimen occurs in a fine-grained limestone, one which fortunately splits easily around the enclosed fossils. This species has a small central disc, and long, slender arms. A related genus, *Astropecten*, occurs in Tertiary rocks, and survives today. Maximum diameter 7 cm.

31 Carpoid, *Cothurnocystis evae*, Ordovician, Ayrshire, Scotland. This curious species is preserved in a fine-grained sandstone as an external mould – that is, all the original shell material has been dissolved away, but this still allows an accurate view of the original organism, especially if an artificial cast is taken from the natural mould. The greatest length of this specimen is 4 cm.

32 Fossil brittle (serpent) star, *Palaeocoma egertoni*, Jurassic, Lias, England. A beautiful species preserved in an impure limestone. The limestone was deposited in very quiet water conditions, which accounts for the preservation of this delicate little fossil. The long slender arms are flexible, unlike the starfish shown previously. Other localities for brittle stars include Miocene rocks in Maryland, USA. Maximum diameter 8 cm.

30   **32**

Permian blastoid, *Deltoblastus delta*. These beautifully preserved little blastoids are stained slightly, because of iron compounds. The slender stems, like those of the crinoids are not preserved here. Compared with the other blastoids in this book the food grooves are very long, extending almost to the base of the animal, and occupying much of its surface area. This species may be one of the last of this kind of echinoderm, which did not survive beyond the Permian. This rare species is unlikely to fall into the hands of the amateur, being from a unique echinoderm fauna in the Permian rocks of Amanoebang Province, Indonesia. The largest specimen here does not greatly exceed 2 cm in length.

special watch for fossils of this kind when hunting in Cambrian localities – it is still perfectly possible to discover a completely new kind of echinoderm!

SEA URCHINS – CLASS ECHINOIDEA (Colour plates 25–29)

Anyone who has waded into the clear blue waters of the Mediterranean without proper footwear may have encountered the protective covering of the sea urchins. Many of the living echinoids are protected by spines, some sharp and breaking off easily into the unwary foot, others stout and clublike; a few groups have taken to burrowing into sediment, and the spines have become small and felt-like. Under the spines there is a test composed of hundreds of calcite plates. Of course there is no stem – the sea urchins are self-propelled. Looking more closely at the arrangement of the calcite plates the five-fold symmetry is still in evidence – in this group there are five areas of finer plates (*ambulacra*) radiating from the centre. Under a lens these small plates show perforations, and during life the tiny tube feet passed through these to assist the animal in locomotion. The perfect polygonal joinery of all the plates is a striking feature of the echinoid test. Some sea urchins are almost spherical, with large strong plates looking rather like shields with a boss in the centre. These kinds of sea urchins carry the stoutest of spines attached to the bosses. Many living sea urchins have undergone a modification of the pentameral symmetry typical of the rest of the echinoderms – they have become bilateral again. Perhaps this is not surprising in animals that are accustomed to moving in a particular direction – they need a front and a back. In burrowing forms the mouth moves to one (forward) side and·the anus to the back, which is obviously a sensible arrangement. Some species have become greatly flattened like the living sand dollar, which has a covering of very fine spines and can bury itself in sand with remarkable speed.

The sea urchins are among the most efficient scavengers in the sea – hence their somewhat unsavoury association with human effluvia. They also include species that eat their way through enormous quantities of sediment, extracting edible particles from it, and in the process reprocessing all the sediment; some of the fossil heart urchins probably had this habit (*see* p. 128). These vacuum cleaners of the sea are vital in preventing fouling of the marine environment. Species with club-shaped spines include active hunters (although they tend to hunt sluggish animals like other sea urchins) and they have powerful jaws on the lower side of the animal that can munch their way through a sand dollar as if it were a biscuit. Some urchins use their jaws to make burrows in solid rock.

Like other major echinoderm groups the geological record of the sea urchins goes back to the Ordovician. In my experience they are generally rare fossils in the Palaeozoic, but as so often with echinoderms there are localities where large numbers of specimens can be recovered from a bedding plane or two. They must have been gregarious from the start. The early sea urchins tend to have rather a large number of plates in a much less regular mosaic than their later relatives. By the end of the Palaeozoic, echinoids with club-shaped spines and beautifully regular tests had become well-established, and their distant descendants survive today. It was in the Mesozoic that a really considerable proliferation of urchins occurred, and they acquired the importance in the marine economy that they retain. They are abundant in Jurassic and Cretaceous rocks, represented both by the spherical spiny forms, and some of those with bilateral symmetry. Isolated spines are common fossils, and useful ones too, because their patterns vary from species to species. Some of the burrowing forms have left their burrows behind as well. In England sea urchins are especially easy to collect from the Cretaceous chalk, where they have been used as one of the fossils for dating the rocks.

STARFISH – CLASS STELLEROIDEA (Colour plates 30, 32)

The starfish are among the most important of invertebrate predators living today. They make major depredations on oyster beds and the recent activities of the 'Crown of Thorns' starfish (*Acanthaster*) in chewing up great chunks of the Great Barrier Reef have become a modern ecological object lesson (and, incidentally, a source of funds for many marine biologists). There are two major kinds of starfish (which may not be closely related) – the familiar large, stiff-armed *asteroids*, and the delicate *ophiuroids*, brittle stars, with lithe, slender, snake-like arms radiating from a circular central disc. Ophiuroids are especially abundant in deep sea environments, and it is unusual for a deep sea dredge to miss one of these small animals. Asteroids often have five arms, but many species have more – up to about forty.

Although they are common today starfish are not generally common fossils, but as so often with echinoderms, they seem to be local in occurrence. Their remains are commoner from the Mesozoic onwards, and at some levels, as in the Cretaceous chalk, they can be found in considerable numbers.

Their history goes back at least to the Ordovician, although some of the Palaeozoic asteroids are distantly related, if at all, to the living species. Ophiuroids have fossil representatives as old as Carboniferous.

THE CARPOIDS – ECHINODERMS OR CHORDATES? (Colour plate 31)

One of the oddest groups of animals covered with calcite plates are known as carpoids (occasionally referred to as calcichordates). Although they have been known for many years their evolutionary position is still hotly debated. They lack any pretence of five-fold symmetry, some are bilaterally symmetrical, but others have no obvious plane of symmetry at all, a most unusual thing for any animal. They appear to have a stem: could they be related to the crinoids? But the 'stem' tapers out to a point, and in some cases had a whip-like flexibility; the same structure has been interpreted as a 'tail'. Similar kinds of arguments can be applied to different parts of these extraordinary fossils. Internally they are really quite complicated, and some possess lobes and channels incised within the plates strikingly similar to the appearance of the brain and nerves in some fishes. Some authorities

Cassiduloid sea urchin, *Pygurus costatus*, Jurassic. The species is preserved without flattening in a hard limestone, which fills the interior of the specimen. This species looks rather like a bun, but not highly convex, and with a five-sided outline. The five ambulacral areas – like the petals of a flower – are clearly visible. They are not sunken into hollows as on many echinoids. This genus and its allies are quite commonly found in Jurassic and Cretaceous rocks. This species is from the Corallian rocks of England, but its relatives occur in Europe, North America and Africa. This specimen has a diameter of 8 cm. Related species often grow to twice the diameter. Jurassic echinoids in this sort of preservation in hard limestones are often difficult to clean. Sometimes the natural processes of erosion will etch a perfect specimen.

have suggested that the carpoids may include ancestors of particular chordates. They have a record going well back into the Cambrian, when it might be supposed that the chordates were undergoing a major diversification. Regardless of whether they include *direct* ancestors or not, the carpoids do serve to demonstrate the likelihood of evolutionary links between chordates – including vertebrates – and the echinoderms.

Carpoids are rare fossils ranging from the Cambrian to the Devonian, when they disappeared completely – unless their descendants live on in one of the chordate groups.

THE ARTHROPODS – PHYLUM ARTHROPODA

There are more living arthropod species than all other phyla combined; if diversity is a measure of success then the arthropods are the easy winners. Arthropods are those animals with an external, usually 'chitinous' skeleton (*exoskeleton*) and have characteristic legs, feelers, etc., with joints to give them flexibility (arthropod is derived from the

Starfish, *Crateraster coombi*, Cretaceous. The starfish is preserved in soft white chalk, which fills in the interior of the animal. The individual calcite plates can be clearly seen. This starfish has a broad central disc, and slender arms. They are stout plates around the perimeter, but tiny-plates in the central part of the animal. Starfish are uncommon fossils, particularly in this state of completeness, since they easily break into their component plates. A few species generally resembling this one occur in Jurassic and Cretaceous rocks. This specimen is from Kent, England, but similar forms occur in the chalk over Europe and in Cretaceous rocks in North America. Related forms survive today. The specimen measures 10 cm across its longest diameter.

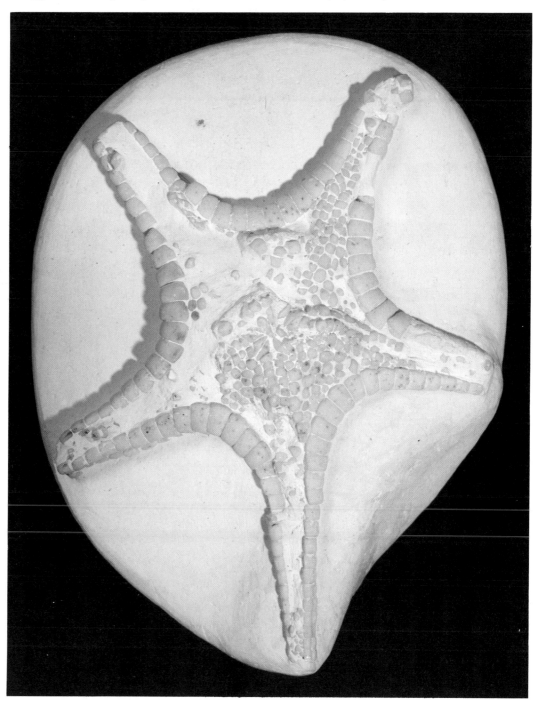

Helicoplacoid – a peculiar and puzzling Cambrian echinoderm

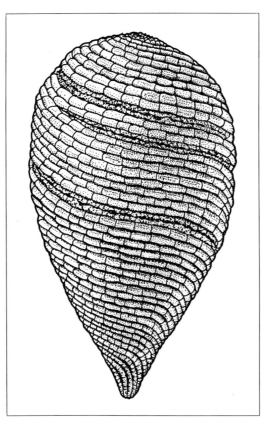

Greek for 'jointed leg'). They have bodies divided into segments, and their legs and jaws are arranged in pairs. They are mostly small and there is a good physiological reason for this: gigantism is mechanically difficult for animals with external skeletons, and above a certain size it is difficult to get enough oxygen into the organism to sustain activity. Despite this one limitation the arthropods have the most varied habits and habitats of any phylum: they can be found anywhere on land, sea or in the air. They inhabit the driest parts of the desert and the deepest parts of the ocean with equal facility. This is partly due to the flexibility of the external skeleton – it can be modified into wings, fortified into claws, sealed against searing heat, or turned into eye lenses. Segmentation, too, has proved very versatile; individual pairs of legs, or *appendages*, can be specialized for particular functions without affecting the routine business of locomotion taken by the others; appendages are modified for feeding in a host of ways, for grasping, swimming, spinning, copulating, cleaning or camouflaging. And their small size enables tiny arthropods, some almost too

Tube of polychaete worm, *Rotularia bognoriensis*, Eocene. Serpulid tubes are preserved here in a siltstone, with parts of the original shell material adhering. The same piece of rock also includes internal moulds of gastropods and bivalves. Worm tubes like these may, at first sight, be confused with gastropod molluscs. The irregularity of the tubes, which become straight near their apertures, distinguishes them from snail shells. Serpulid worm tubes are found in many different sediment types in rocks of Mesozoic and Tertiary age. Some species are very tiny, and encrust other shells. This species is from Bognor, England, but related forms occur the world over. Overall length of specimens about 3 cm.

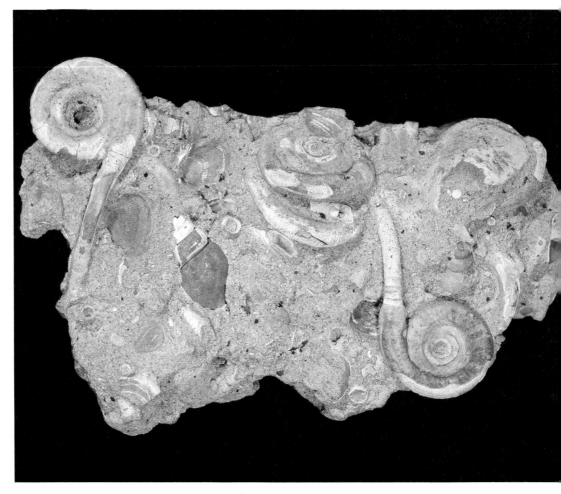

small to see with the naked eye, to live in crannies, within the soil, between sand grains, and the like. They are truly ubiquitous.

They are also an ancient group, since obvious arthropods, of primitive kinds, can be found in the earliest of Cambrian rocks. At this time, of course, they were wholly marine. They were probably derived from some sort of segmented worm (annelid) – a few scientists adopt the view that different arthropods may have come from different and separate worms, independently, which became 'arthropodized' by acquiring an external skeleton. With certain exceptions, the fossil record of the arthropods is not commensurate with their diversity today, and their probable diversity in the past. Some forms have delicate exterior skeletons that are only preserved under exceptional circumstances (*see* p. 16). Many terrestrial forms simply did not live in environments in which preservation was possible. So the record of the group as a whole is a bit patchy, with 'jumps' in time between one fossil find and the next. Some groups, those with good, solid (and especially calcified) hard parts are

frequent and important fossils, with an unparalleled record. A brief summary of the arthropods cannot do justice to all the fascinating specialized groups – sea spiders and mites for example – that have an incomplete fossil record. We shall concentrate on the arthropods with the best fossil record. Some of these are common enough to dominate fossil assemblages, and they are among the most attractive of the common invertebrate groups.

TRILOBITES – CLASS TRILOBITA (Colour plates 33, 34, 36, 43)

The North American Indians had a name for trilobites that meant 'little water bug in the rocks', which shows extraordinary zoological acuity. The trilobites were one of the dominant forms of marine life in the Lower Palaeozoic, known from thousands of different species and found in every continent. Not surprisingly they were among the first fossils to attract widespread attention. They have a high degree of organization and a variety of form that shows how quickly the arthropods were able to exploit the advantages of their external skeleton. All trilobites

Cambrian trilobite, *Wanneria walcottana*, preserved on a fine sandstone, with a certain amount of crushing. This is one of the most primitive trilobites known. The crescent-shaped eyes are attached to the middle part of the head. There are a number of similar species from early Cambrian rocks distinguished on the size of the eyes and details of the thorax. The specimen has a length of 10 cm.

are divided lengthways into three lobes — hence their name. They can also be divided crossways into three regions, a well-defined head (*cephalon*) at the front, usually equipped with a pair of eyes, a thorax with a variable number of articulated segments, and at the back end a tail (or *pygidium*) formed by the fusion of several segments. The parts of the animal we find fossilized are usually only the carapace, the hard parts that formed a protective covering on the back of the animal. If we could have turned over a living trilobite we would have seen an array of jointed legs on the underside, and flexible antennae. The traces of these appendages *are* preserved in exceptional circumstances, so we do know that trilobites had walking legs, gills, antennae, and the bases of the legs modified into primitive jaws in some species. The appendage pairs are generally similar along the length of animal. Few examples of complete trilobites are known and for the most part we have to be content with the exterior skeleton.

The great variety of form of the hard parts surely shows that trilobites were adapted to a wide range of habitats and life styles. We know that they could live in extremely shallow to very deep-water environments, that most lived on or near the sea bottom, but a few were adapted to open-ocean swimming. They probably had a variety of different habits, some no doubt were scavengers, others sediment grazers, and it seems probable that some trilobites were also active hunters, pursuing worms and other soft-bodied prey, which they shredded using the spines on the base of their legs. The eyes of trilobites include some very sophisticated structures that show they were far from primitive visually. Many species have thousands of lenses — like a fly's eye —

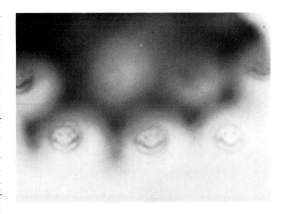

Photograph of 'happy buttons' taken through the lenses of the eye of the trilobite *Phacops* (*see* p. 87) of Devonian age.

Cambrian trilobite, *Elrathia kingi*. This trilobite has been preserved in a curious way: the original shell has been thickened by mineral growth, still of calcite, to give the fossil a distinctive, medallion-like appearance. This trilobite has a large number of wide thoracic segments, not greatly convex. The tail is wide, well-segmented. The inflated middle part of the head tapers forwards. This species occurs in great numbers in the Wheeler Shale of Utah, USA. Related species occur over a much wider area around the margin of North America and in Europe. Complete specimens are usually $\frac{1}{2}$–3 cm long. They are sold in rock shops and souvenir centres, often mounted into jewellery.

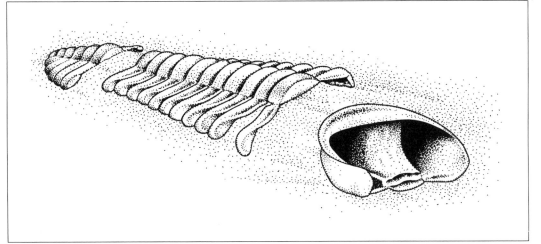

Moulting in trilobites; breakup of the exoskeleton into several pieces helped the animal shed its old carapace

Trinucleid trilobite, *Trinucleus fimbriatus*, Ordovician. An extraordinary little trilobite looking like an ancient medallion, preserved in shale, with some of its original *exoskeleton* retained, coloured by iron compounds. Trinucleid trilobites are blind, and with a greatly inflated mid-part of the head region. Around the perimeter of the head there is a pitted fringe, which is unique to this kind of arthropod. Trinucleid trilobites, with various types of fringe, are characteristic of Ordovician rocks. There are more than 100 species. This species is from Builth, Wales. While perhaps commonest in Europe (Wales, Scotland, Sweden, Norway, Czechoslovakia), trinucleids are also known from North America (Virginia, Nevada and various localities in Canada), North Africa and South America. Specimens do not usually exceed 2½ cm in length.

which were very sensitive to detecting any movements in their field of view. Others had fewer lenses of remarkably efficient construction; photographs have actually been taken (p. 86) using these ancient lenses in a camera – and they still focus precisely after 400 million years or so. The lenses, like the rest of the exoskeleton, are made of calcite.

Trilobites vary in length from a millimetre or two to fifty centimetres or more – the 'average' trilobite is 5 cm or so long. It is actually rather unusual to find a complete trilobite as a fossil – usually they are in pieces. Trilobites, like most arthropods, grew by moulting (casting off the old carapace and growing a new, larger one) and many of the bits we find are probably the 'cast offs' and not the dead animal. During moulting the head split into several pieces along special *sutures*, the thorax often parted into its individual segments and the tail was displaced (*see* p. 86). The early moults, 'baby trilobites', have been discovered. Some trilobites lost their eyes, probably those that burrowed in mud or lived in lightless parts of the ocean. Many trilobites could enroll when threatened, just like the living woodlouse – some of them

even evolved locking devices to make their enrollment really secure. In some places you can find dozens of enrolled trilobites together; these are the remains of the animals themselves, not the moults, which presumably perished together after a fruitless attempt to protect themselves from a miniature catastrophe such as a sudden influx of sediment.

Typical trilobites of several kinds are present in the earliest Cambrian rocks; they must have had late Precambrian ancestors, but of these nothing is known. They swarmed in the Cambrian seas, outnumbering all other invertebrates in most localities. The trilobites evolved rapidly, and can therefore be used extensively in the dating of rocks at this time. They continued to proliferate in the Ordovician, although many of the Cambrian kinds had died out; in terms of the variety of forms present this period was perhaps their heyday. Several kinds of trilobite failed to survive the end of the Ordovician, but other trilobites were still conspicuous during the Silurian and Devonian. Only a few major groups survived into the Carboniferous, but they can still be lo-

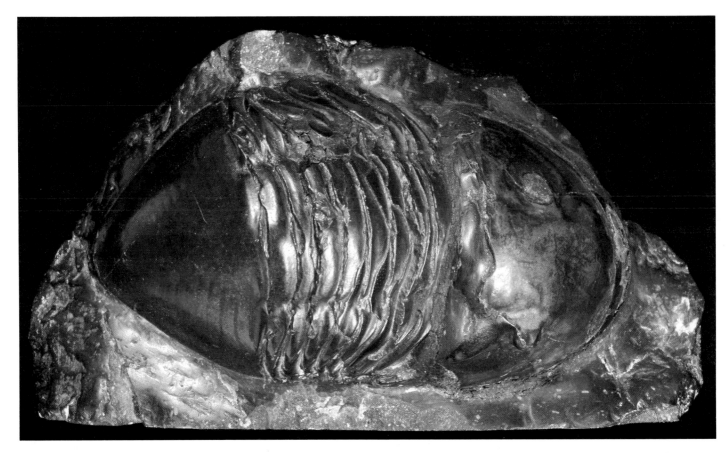

cally abundant. The last trilobites are found in Permian rocks, and before the Triassic they had disappeared forever. We might perhaps hope that one species lingered on, like the brachiopod *Lingula*, in some inaccessible part of the sea, but as even the deep sea faunas become well-known this possibility is fading.

Since many of the extremely ancient rocks in which trilobites can be found are folded or otherwise distorted their remains can often be found twisted or fractured. But in other sites, particularly in limestones, they can be found with every fine skeletal detail preserved. The many different kinds are distinguished by the nature of the segmentation, particularly in the head region, the structure of the thorax and pygidium, the size, position and structure of the eyes, the development of the moulting sutures, and the surface patterns on the exterior of the exoskeleton. New trilobites are being discovered every year, and there is always the possibility of finding one with its legs preserved!

CRABS, LOBSTERS, SHRIMPS AND THEIR ALLIES – CLASS CRUSTACEA (Colour plates 39, 41)

The seas today swarm with crustacean arthropods ranging in size from the tiniest krill in the surface waters to giant bottom-living crabs with a reach like that of a man. The majority of crustacea have appendages that are specialized to do different jobs along the length of the animal – some for grasping, some for swimming and so on. Most smaller crustacea have exoskeletons that are rather delicate, and for this reason the fossil record of the shrimps, for example, is somewhat incomplete. Undoubted crustaceans are found in rocks as old as Cambrian, at which time free-swimming species were living happily alongside the trilobites. It is not until the Carboniferous that we start to find evidence of a variety of forms including the putative ancestors of some of the living dominant groups. Crabs and lobsters have the thickest exoskeleton of all the group – one like that of the trilobites reinforced with calcium carbonate – and the fossil record of this important group (only a small part of the whole class however) is much better. Lobster-like animals are found as far back as the Triassic, and both crabs and lobsters are frequent and appealing fossils in Cretaceous and Tertiary rocks. The group was hardly set back at all by the major extinction at the end of the Cretaceous, and has never been more varied than it is today. The most highly

Asaphid trilobite, *Isotelus gigas*, Ordovician. This fine, large trilobite is preserved in relief in a limestone – a common mode of preservation for North American Ordovician trilobites. A 'smoothed out' trilobite, with the head of similar size and shape to the tail, and only eight thoracic segments (contrast the Cambrian trilobites in this book). The eyes are small. Asaphid trilobites of many species are common in rocks of Ordovician age. Many grew to an impressive size. *Isotelus* is typical of the North American continent, where it occurs in many localities. Other asaphids occur in most countries of Europe, especially Norway and Sweden, and widely in Asia and Australia. This specimen has a length of 9 cm.

specialized crustacea are probably the barnacles (Cirripedia) and individual barnacle plates are rather commonly encountered in rocks of Cretaceous age and younger. Finally the diminutive, bivalved ostracods are important and abundant fossils, which we will consider with other microfossils in Chapter 8.

SPIDERS, SCORPIONS, SEA SCORPIONS, HORSESHOE CRABS – CLASS CHELICERATA
(Colour plates 37, 38, 40)

This large and varied group of arthropods includes the giants of the phylum, as well as tiny spiders hardly visible to the unaided eye. They are classified together because of basic similarities in the appendages, even though they have diversified into very different-looking creatures living equally happily on land and in the sea. Predatory habits are typical, and many species have specialized clutching appendages, or stinging organs. None of the group is especially common as fossils, but they make up for their general rarity by their interest. Most spectacular are the sea scorpions (eurypterids), which include arthropods as large as any that have lived (two metres or more in length). Appendages in the 'head' region often include powerful claws, and it does not take much imagination to conclude that these eurypterids were among the fiercest predators of their day. They were prominent in the Silurian and Devonian, where they can be found with the fish-like animals that abounded in the fresh- and brackish water deposits of the time, although some of the Ordovician ones are in marine sediments. Some of the eurypterids may even have made tentative excursions on to the land. They are also probably the earliest animals in which two sexes can be certainly identified. The eurypterids are probably allied to the scorpions, which originated in the Silurian or even earlier, and successfully made the transition from the aqueous environment to land. They are abundant enough today in warmer climates to be a serious cause of death in some countries. Perhaps it was the transition to land that enabled the scorpions to survive beyond the Palaeozoic, which saw the end of the sea scorpions. Spiders are known as far back as the Carboniferous, but their remains are principally known fossilized from the Tertiary ambers, where perfectly preserved specimens retain even the hairs on the legs. The Horseshoe Crab (*Limulus*) is one of the most celebrated of 'living fossils', and it is a genuine survivor having changed little in 150 million years (*see* p. 91). It is the last of the *merostomes* – the group of fossil horseshoe crabs that were varied and numerous in the coal swamps of the Carboniferous and have a history that extends back to the Cambrian. Why *Limulus* alone should have survived is something of a mystery. It appears to have rather generalized habits, feeding on everything from worms to clams, which it can crush with its powerful appendages, using the bases of its legs like nutcrackers. It can swim, flipped over on to its boat-like carapace, or crawl. Presumably its ancestors were not so different when they shared the marine habitat with the trilobites.

Below left Fossil spider, *Eomysmena moritura*, Oligocene. This is another amber specimen, exquisitely preserved in its entirety. It is an important and rare specimen. This species has a short and fat abdomen, which preserves something of its original light colour. The legs are shorter than those of the other spider in this book, and the four pairs are clearly visible. This fossil specimen is probably unique. There are species living today related to this fossil species. This specimen was obtained from the Baltic bays. The spider is just over ½ cm long

Below Fossil swimming crab, *Notopocorystes broderipi*, Cretaceous. This crab is preserved in a rather soft clay, which makes it more fragile than the other crabs in this book. The crab has a longer body than the other species shown here, and the legs are flattened and held away from the carapace. There are a number of species generally similar to this one in rocks of Cretaceous and younger age. This specimen is from Folkestone, England. Length of the carapace is 2½ cm, but the species grows to more than twice this size.

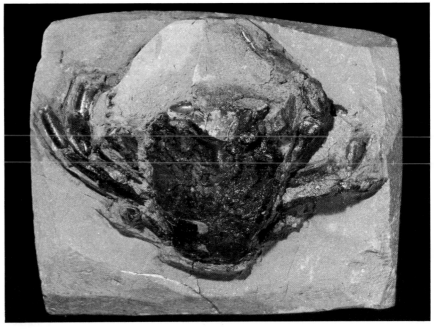

89

The insects – flies, butterflies, beetles, ants, fleas, bugs, dragonflies, grasshoppers, aphids, wasps, etc. – are the most varied and numerous of the arthropods both in numbers of species and individuals. They are also the most perfectly adapted of terrestrial organisms, inhibited neither by extremes of climate nor altitude. If this book were written in proportion to the number of species in the group the rest of it would have to be devoted to the insects. But the fossil record of the insects is far from perfect – very sporadic and selective. Where insects *are* preserved they seem to occur in large numbers and in variety, so that a few famous localities have yielded a disproportionate number of the fossil species. Insects are fragile and easily damaged, and it is not surprising that they are sometimes destroyed completely. Many species live, and presumably lived in the past, in inland or upland sites where little sediment accumulates. Naturally the commonest insect fossils are the most durable parts, like the wing cases (*elytra*) of the beetles. But their small size also means that insect fossils are easily overlooked and they may be more common than is supposed. The best sites are in rocks originating from sediments that accumulated in quiet, freshwater sites – lake bottoms, peat bogs and the like – and they are generally rarer in marine facies.

Most insects have six legs in three pairs, and are equipped with wings (a number of primitive forms are wingless). The early evolutionary stages of development of wings are still in question – they may have developed from outgrowths of exoskeleton from the body that assisted gliding initially, and then acquired a propulsive function. In any case winged insects were well established by the Carboniferous, when some of the giant 'dragonflies' (not really a close relative of the living species) had a wingspan of tens of centimetres. No doubt many of the crucial steps in insect evolution took place during the Devonian, but of this phase of insect history there is very little record. In the Carboniferous and Permian the insects were highly diverse, and there is still debate over how closely these forms are related to our living fauna: some forms, like the cockroaches, undoubtedly have a long history. Flight gave the insects the chance to exploit habitats previously vacant, and unrivalled dispersal. The extraordinary flexibility of the group has meant singular problems in their classification, but certain characteristics, like the structure of the mouthparts or the wing vein patterns serve to define major subdivisions. During the Mesozoic some of the most important types of insects still liv-

Fossil crab, *Xanthopsis rugosus*, Eocene. This specimen comes from a clay; the best specimens often occur in slightly harder, nodular layers within the clay. Two strong furrows run across the back of the carapace of this crab, which is knobbly, and, on a fine scale, covered with minute pits. As with *Palaeocarpilius* powerful claws are held close to the body, like the living shore crab. A number of *Xanthopsis* species are one of the characteristic crabs of Eocene deposits. This specimen was obtained from a tunnel under London! Similar specimens are to be obtained from elsewhere in southern England, in France, and in the USA. Diameter of the carapace 6 cm.

The living horseshoe crab, *Limulus*; compare with Fig. 38.

ing – like the butterflies – originated. Butterflies have an intimate relationship with flowering plants, which they pollinate, and themselves derive nourishment from the nectar. It is difficult to imagine a butterfly without a plant to feed on, and we have to assume that both the plants and the appropriate pollinators diversified together. Other kinds of insects besides butterflies have used the nectar of plants for a livelihood, including bees and some flies.

The fossil record of insects is at its best from Tertiary deposits, where amber, the fossil resin of coniferous trees in which insects became entrapped, yield great numbers of exquisitely preserved specimens. Even the most delicate are present. The oldest amber is Cretaceous. The fossil remains of beetles, including the hard elytra, have been extensively used in the interpretation of the climatic history of the Pleistocene ice advances and retreats, because the species of beetle present in peat deposits change in harmony with the climatic fluctuations.

PHYLUM CHORDATA

The chordates include the dominant living animals, if you measure dominance by size, or the capacity to alter the world. Man is a chordate. The phylum at the present is dominated by its more advanced members – the vertebrates. But in the Cambrian or earlier the chordates no doubt included animals little more highly organized than the contemporary arthropods or brachiopods. A few of these survive. The lancelet, *Branchiostoma*, one of the standard laboratory animals for dissection, is a primitive chordate (there is a single fossil lancelet from the Middle Cambrian). Hardly anything is known about the fossils from the earliest history of the phylum, possibly because the early chordates lacked hard parts, unless the enigmatic carpoids belong here. The acquisition of a *skeleton* of calcium phosphate marked the point at which the fossil record began to contribute to the history of the group which, from our selfish viewpoint, we are conditioned to think of lying at the pinnacle of all evolutionary trees. Curiously enough the use of phosphatic material to build hard parts is generally primitive – calcium phosphate shells are nowhere more numerous than in the earliest Cambrian. All chordates have a nerve chord running along the back, and all higher chordates were derived from forms with gill slits. We are mostly concerned with the history of the vertebrates. Once the vertebrates were established they quickly acquired hard parts displaying great complexity when compared with most invertebrate phyla (we should except the arthropods, perhaps). So it is possible to deduce more about the course of evolution from the study of vertebrate skeletal remains than is possible with most invertebrate groups, but even now, many mysteries remain.

As a generality, the fossil record of the vertebrates is not as complete as that of many of the invertebrate groups, and particularly those with a continuous fossil record like the ammonoids. Vertebrate remains are decidedly patchy, particularly the terrestrial species. This is not altogether surprising considering the factors militating against their preservation – break-up of skeletons after death, the necessity to have a skeleton incorporated with sediment, and the activities of predatory and scavenging animals to destroy remains. So land vertebrates in particular tend to come from relatively few sites, which become exhaustively collected compared with most invertebrate localities. Some of the sites are enormously rich, however, and the history of palaeontology is punctuated by quite unscientific feuds between experts trying to find and hold on to the best sites for the most spectacular vertebrates. The more impressive the animal the greater the feud; the fights between Professors Cope and Marsh in the last century over the dinosaur remains of North America include examples of double dealing that would not shame an oil tycoon. Even fish remains are rather local, and do not adequately reflect the abundance of a group that has occupied the seas for more than 400 million years.

There is room here to give only the briefest account of the different kinds of vertebrates. The higher vertebrates, fossil mammals and dinosaurs in particular, have been described in so many books that they sometimes seem to have as much flesh and blood as any animal in the zoo. The dinosaurs alone have commanded as much popular attention as the rest of the fossil animal kingdom combined. This reflects their spectacular size and the drama of reconstructions of battles to the death between *Tyrannosaurus* and its armoured contemporaries. The sad truth is that the average collector is unlikely to recover more than the most fragmentary remains of dinosaurs. The commonest vertebrate remains are of teeth or other resistant pieces. Fortunately the teeth of higher vertebrates can tell us a good deal.

JAWLESS FISH (Colour plate 47)

The oldest fossil fish, which are the earliest known vertebrates, lack jaws – the mouth is a simple opening – and are covered on the *outside* with bony skeleton. The first ones seem to have lived in the sea, but by the Silurian the jawless (Agnathan) fish were a prominent component of fresh- and brackish-water sites in many areas of the world, and continued to diversify into the Devonian period. Some of these freshwater fish are heavily armoured and make robust fossils – particularly the headshields. The insides of the shield have been known to preserve quite tiny details of the nervous system, for example. It is probable that the jawless fish lived by grubbing in the sediment, or perhaps by filter feeding, possibly exploiting organic material derived from the plants that were taking to life on land at the same time. A few living jawless fish are the only remnant of this ancient group, and they are highly specialized forms. The hagfish feeds on dying fish using its rasp-like tongue in conjunction with a solitary tooth on the roof of its mouth. The lampreys (*Petromyzon*) are rather nasty external parasites of other fish, to which they attach themselves with a sucker, and proceed to rasp away at the living flesh. The specialized habits of both these 'pseudo-fishes' may have contributed to their survival, when none of their bony-plated relatives survived the Palaeozoic. Modern specialists often divide the jawless fishes into two broad groups, one containing the hagfish and the other the lamprey, with the fossil jawless forms distributed between the two.

Jawless fish are particularly common in red rocks of the Silurian and Devonian periods, which was a time when non-marine rocks were widespread, and have survived subsequent erosion. They accompany the eurypterid arthropods (which may have preyed upon them), and we can visualize the rivers and lakes of the time thronging with the first invaders from the sea. The subsequent development of the jaw enabled the vertebrates to exploit a wider range of lifestyles than were available to the jawless fish, and it would not be too gross a generalization to say that much of the evolution of the vertebrates was intimately related to things that happened to jaws.

SHARKS AND ALLIES

The sharks and their relatives (known by the indigestible class name Elasmobranchiomorphi) are the most successful of long survivors among the vertebrates. Their rapacious habits have been the subject of bloody hyperbole in films and bestsellers, so everybody knows that the shark has ranks of

The early jawed fish, *Cheiracanthus murchisoni*, from the Devonian. The fish is preserved in a siltstone nodule, which has partially formed around the body of the fish. The body is of normal 'fishy' outline covered with minute rhomboidal scales. Note prominent spines which supported paired fins. The species is named after Sir Roderick Murchison, a great pioneer of the study of Palaeozic rocks. Acanthodian fishes generally resmbling this one are known from quite complete specimens from Devonian to Permian rocks. Some achieved a considerable size. In the Devonian rocks, fishes of this kind are known from (mostly) freshwater deposits in Scotland, Wales, Norway and arctic Canada. This specimen is from Banffshire, Scotland. It is 15 cm long.

fearsome, pointed teeth, an elegant, incessant swishing swimming motion, a tail with a long 'point' uppermost, and a big appetite. Their skeleton is made of cartilage – bone has not 'grown into' the cartilage, as in the higher vertebrates. Their true jaws are of obvious advantage for grasping prey. Cartilage does not preserve as fossil as a rule, so most of the evidence of the shark-like fish rests upon teeth. Sharks of essentially modern type go back to the Jurassic. In the later Palaeozoic a number of shark-like fish include some related to living forms, others whose zoological position is still a puzzle. The Devonian *placoderms* are an especially exciting group of fish (*see* p. 94) with powerful, armour-plated heads and frequently fang-like teeth, the function of which is obviously predatory. The lower jaw of these fish is loosely slung to the brain case so that they had a very wide 'gape'. The trunk was also protected with a shield. This group may include the distant ancestors of the sharks, but they must have been somewhat ponderous animals compared with the streamlined hunters of modern seas. Most of them were dogfish-sized, but *Dunkleosteus* was twice the length of a man, and was surely the most formidable predator of its day.

Sharks of modern type replace their worn teeth with new ones by a sort of conveyor belt system. Isolated, bladed shark's teeth are really rather frequent finds in Mesozoic and Tertiary rocks, often in the absence of any other fish remains. They have a hand-some shiny lustre which makes them conspicuous from a distance. The rays are related to sharks, and are in essence sharks that have been 'squashed' flat in response to life as rather sluggish bottom dwellers. The mollusc-crushing teeth of rays, with the same lustre as shark's teeth, but formed as low, ridged cushions, can also be found in Mesozoic and younger rocks.

BONY FISH – OSTEICHTHYES (Colour plates 46, 48, 49)

Every experience with an unfilleted kipper demonstrates the aptness of the general name of this group. The vast majority of living fish belong here (18,000 species or so), which rival the insects in their multitude of adaptations and variety of external form. They can live in hot springs, or can be dredged from the deepest abyssal depths of the ocean; the latter are grotesque gargoyles that seem to belong in the paintings of Heironymus Bosch. Some are rapid swimmers gathered together in silvery shoals, others sluggish bottom dwellers spending most of their time buried in sediment. Most have a characteristic arrangement of paired fins, but these can be lost, or fantastically modified to form fans or poisonous barbs. Most species are a few centimetres to a few tens of centimetres long. Their rapidity of evolution is well known: dozens of species have evolved in African fresh water lakes within the last million years. They have as wide a range of life habits as any group, from scavengers, grazers, filter feeders, vegetarians, omnivores, to fierce predators with reputations to rival that of the sharks. A coating of *scales* is characteristic, and the gills are covered by a movable flap (the *operculum*).

It would be difficult to overestimate the importance of the bony fish in the economy of the sea. Fish of various sizes feed on the plankton (and on each other), and themselves are food for larger predators, such as seals and many whales, and man. Many of them have bizarre adaptations, as parasites, or as permanent 'guests' within particular species of coral or sponge, which it would be impossible to infer from fossil remains. Many have good sight, but there are a few forms which have lost their eyes, some of them living in the lightless world of caves.

Considering their abundance in the oceans and rivers today, bony fish fossils are rather rare, and poorly reflect their true importance. Where they occur, however, they are found in large numbers, and in variety. Presumably these sites were protected from the kinds of influence that normally destroy fish remains, such as the activity of scavenging animals that disarticulate the skeletons, or currents. The classic fish sites produce fossils of quite outstanding beauty (*see* Plate 49) and ones which have considerable monetary value. The fish of the Eocene and later are clearly related to today's fauna, and some relatives of the living groups can be found in Mesozoic rocks. The classification of bony fish is extremely complicated, particularly fitting in the fossil forms, and there are many bones of contention (literally!). The group as a whole goes back to the Devonian (?Silurian) period, and from rocks of this age predatory fish (*Palaeoniscus* and allies) have been recovered. Some living survivors of these early forms include the sturgeons and bichirs (the former have reverted to a cartilage skeleton). The early fossil bony fish have tails with the long blade uppermost, whereas the more advanced living bony fish have a tail that is symmetrical, and more effective in producing a horizontal thrust. The surface of their body was covered with an interlocking series of shiny scales ('ganoid') which are still found in the bichir.

One important division of bony fishes includes the lungfish and the 'lobe-finned' fishes (including the coelacanth, the most famous of living fossils). This group of fish was also fully fledged in the Devonian, and this is of particular significance because the origin of all land vertebrates has been considered as lying within a species of this

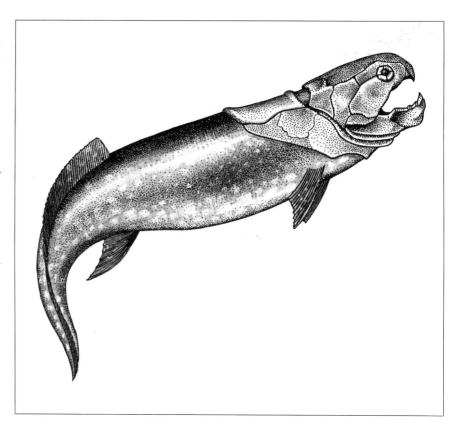

Armoured placoderm fish of the Devonian

group. The freshwater lungfish retain their lungs, which help them to survive periods of drying in the rivers in which they live. Many fish specialists believe that in the Devonian all the bony fishes had lungs (and that the swim bladder of the recent forms was a modification of the early breathing apparatus). It is not an easy business to identify the actual ancestor of the land-going organisms, and perhaps it has not yet been found. For many years there have been proponents of the theory that one of the 'lobe-finned' fishes included the ancestor of the tetrapods, and it is not difficult to imagine their stumpy fins, with a fleshy core, making the uneasy transition into a walking leg. Other features do not agree with this, and recently there has been a swing to favour the lungfishes as nearer the origin of amphibians and, ultimately, ourselves. Whatever momentous steps they took in the Devonian, the lungfish and lobe fins have been very conservative since, for the living lungfish are clearly similar to their Palaeozoic relatives and coelacanth lobe fins like the living genus (*Latimeria*) are known from Cretaceous rocks.

AMPHIBIANS – CLASS AMPHIBIA (Colour plate 51)

Frogs, newts, and salamanders are living amphibians. The frogs in particular are one

Bony fish, *Lycoptera middendorfi*, Cretaceous. This little fish is preserved in a soft, pale grey shale. Like many shale specimens it is flattened, but the still-water conditions under which the rock was deposited have permitted complete preservation of a delicate skeleton. This fish has a rather expanded, frog-like head behind which there are prominent fins; a slender body, which is flexible, supported by a host of tiny ribs, which are clearly visible. A number of related fossil species are known from Cretaceous rocks. This species is from Cretaceous fish beds in Turga, Nertchinsk, Siberia. Similar species occur in Mongolia and China. It is 6 cm long.

of the most diverse of the higher vertebrate groups and peculiarly endearing animals with their lugubrious expressions and vocabulary of squeaks and belches. Yet the amphibia as a whole represent an early level of terrestrial organization; they are tied to water for reproduction, because their larvae have to pass through a wholly subaqueous phase, and dessication of the adult is a constant danger. The amphibians had their origins in one of the freshwater lung-carrying fishes in the Devonian. We have to visualize the process of change as a gradual one, with progressive forays from the water, until the first animal with true limbs developed (the bony elements of the limb are present in the fish fin). From the Devonian of Greenland and other localities comes a true four-legged animal (tetrapod) that is not far from the 'fish with legs' we might have visualized (which is commemorated in its name – *Ichthyostega*). But as always it is not quite that simple. The transition from water to land required not only the perfection of a walking limb, but modifications to the hearing system, palate, eyes and so on, and *Ichthyostega* had already established most of these terrestrial modifications, so it is in no sense an 'ancestor', although it has many ancestral features. Bony scales were in-

herited from its ancestors and no doubt this helped the animal against dessication. The subsequent history of the amphibians includes a further development of terrestrial forms, but others remained wholly or partly aqueous. These flourished in the water-logged swamps of the Carboniferous period. The earliest amphibians were certainly predators, and there was suitable prey both on land and in the water.

The Carboniferous is popularly known as the 'Age of Amphibians', and it is true that the amphibians reached a dominance in the vertebrate world at this time which they never again equalled. All of the walking species had the widely-splayed legs that gave them a slow and lumbering gait, but, in the absence of more streamlined animals, they prospered. The group informally known as *labyrinthodonts* (from their characteristic labyrinth-ridged teeth) included some impressive animals almost as long as a man: perhaps the alligators of the time. Most of the fossils recovered are of wholly or partly aquatic amphibians, the sort that were easily preserved in the coal swamps that were characteristic of the Carboniferous non-marine environment. Other environments, particularly later in the Permian period, which was one of general 'drying out', probably included species that were more truly terrestrial in their habits. But even during the Carboniferous the first true reptiles had evolved, and this was the group that was to make itself wholly independent of the water. The living amphibians have turned the limitations of the aqueous connection to advantage. Frogs have their origins in the Triassic (they are a relatively recent amphibian innovation, and highly specialized). Their tadpoles can exploit bodies of water not excessively populated with competitors, and some are adapted to very restricted niches. The adult frogs have a different diet (worms, flies, etc.), and the species therefore get 'the best of both worlds'. Frogs throng in almost any site with high humidity and standing pools of water. The large amphibians of the Palaeozoic did not survive beyond the Permian, and so our inferences about their modes of life have to be made entirely from the bony fossils that survive. Palaeontologists recognize fossil Palaeozoic amphibians from their teeth, their flattened skulls and the arrangements of bones on the skulls. They are not generally common fossils, but there is always a chance of turning up a skeleton from the coal deposits, or finding a fossil frog (*see* Fig. 51) in the Mesozoic.

REPTILES – CLASS REPTILIA

The living reptiles are almost all predators, and the vast majority are also small, the only giants remaining being the crocodiles, alligators and caymans (and one or two big lizards). We tend to think of the reptiles as somehow past their 'prime', but it would be more accurate to say that they had been displaced from the top jobs in nature, while more than holding their own in the shop floor. The warmer the climate the more they are in evidence, because the living reptiles are 'cold blooded' (without their own internal heat regulator) and they cannot successfully live in very cold climates. Living snakes and lizards are highly varied, and manage to live in some environments (e.g. dry deserts) where the mammals are pushed to survive. Most living reptiles and all of the primitive ones lay eggs, and this *amniote* egg, from which perfectly-formed baby reptiles hatch, marks the complete emancipation from the necessity of returning to water for reproduction. And a scaly skin solves the problem of drying out even in intense heat. Add to this the change in orientation of the legs in many reptiles, which can be used in an efficient running action, unlike the ungainly

waddling of the amphibians and it will be apparent why the reptiles were better adapted to terrestrial life than the amphibians, and why they largely displaced them. This was far from instantaneous, and the Permian period saw both advanced amphibians and early reptiles living side by side.

The reptiles were probably derived from an amphibious ancestor, probably early in the Carboniferous. Once established they underwent a number of evolutionary 'bursts' in which diverse kinds of reptiles occupied a variety of habitats, the most spectacular of which was the dinosaur radiation in the Mesozoic, to which we shall return later. By no means *all* of the large reptiles that are found in the Mesozoic rocks are dinosaurs – the reptile groups that took up life in the sea or the air were only distantly related. One of the most important steps for the complete colonization of the land was the evolution of *herbivorous* (plant-eating) reptiles, tapping a prolific and nourishing source of food that opened up a new dimension to the 'food chain'. Larger and better herbivores meant larger and more ferocious carnivores to prey upon them, scavengers to clean up the mess, and a whole host of subordinate trades for other animals. The reptiles underwent tremendous changes in their skeletal structure, their jaws and their teeth during over their history of more than two hundred million years, some of the predators developing razor sharp rows of fangs, other herbivores losing their teeth and acquiring beak-like jaws. At some time in the Permo-Triassic one group of reptiles ('mammal-like reptiles') actually gave rise to the warm-blooded mammals, which were to lead a rather subordinate existence during the heyday of the dinosaurs. And the birds probably had their origin in a particular group of dinosaurs. So in a sense the reptile dominance is still with us, transmuted by time and evolution.

The scenes of Jurassic or Cretaceous landscapes in picture books show dinosaurs swarming over the landscape, giving the impression that we know the whole story. The truth is that there *are* a few sites (especially in North America) where abundant remains of complete animals have been recovered, and from which a passable idea of the fauna of the Mesozoic can be obtained, but there are still great gaps in our knowledge, and many of the dinosaurs are known from a handful of individuals, or even a single specimen. There are doubtless still new kinds to discover.

33 Trilobite, *Calymene blumenbachii*, Silurian Worcestershire, England. This trilobite is exceptionally well-preserved, retaining its original convexity and shell-material. A convex trilobite with the lobation of the middle part of the head distinctly defined. The thorax has thirteen segments. 8 cm long.

34 Giant trilobite, *Paradoxides 'spinosus'*, Cambrian, Bohemia. This large trilobite is preserved in a black mudstone, slightly crushed. The swollen mid-part of the head expands forwards. There are many thoracic segments extended into spines, and a tiny tail. There are a number of *Paradoxides* species, which include some of the largest trilobites. This species is relatively small for the genus (15–20 cm); some species approach 60 cm in length.

35 Fossil rhagionid fly *Chrysopilus* sp., Oligocene. This insect is exquisitely preserved in amber. The amber was at one time polished to make a pendant with the fly as centre piece. This is a Baltic amber specimen from East Prussia. Living *Chrysopilus* live in marshy areas in Europe, their larvae growing up beneath the soil. The fossil is just over 1 cm in length.

36 Devonian trilobite, *Phacops rana*, Ontario, Canada. This species is beautifully preserved in limestone in full relief. With luck the amateur collector should be able to obtain specimens as splendid. A convex trilobite with eleven thoracic segments. The mid-part of the head is covered with coarse tubercles and expands forwards; the eyes include a few, very large lenses. Complete specimens are usually in the range 3–8 cm.

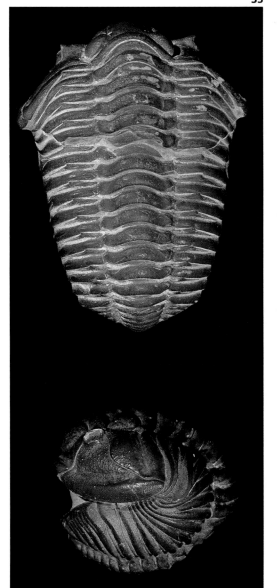

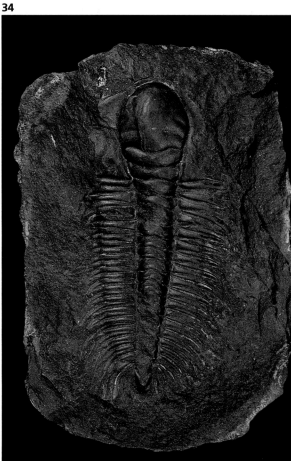

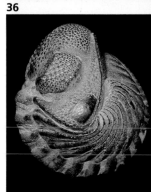

34

37   **38**

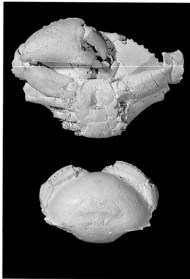

39

40

37 Sea scorpion (eurypterid) *Eurypterus lacustris*, Devonian. The sea scorpion is preserved on a fine siltstone, and is exceptionally complete. Fragments are more usual. The rather fat body, with distinct segmentation, and a long tail-spine are characteristic; the head region carries appendages of grasping aspect. This species is from Canada, but eurypterids occur on most continents.

38 Fossil horseshoe crab *Mesolimulus ornatus*, Jurassic, Solenhofen, Germany. The horseshoe crab is preserved on a flat-bedded limestone. Some of the original flexible material of the skeleton still remains. The king crabs have a nearly circular carapace, beneath which powerful legs helped the animal to swim and catch prey. Note the long pointed tail spine. This specimen is 8 cm long. Larger fossil examples are known.

39 Fossil crab, *Palaeocarpilius aquilinus*, Miocene, Libya. This perfect and nearly complete crab is exceptional. The white colour is because of the white limestone matrix. This beautiful crab has a wide carapace, with a few faint depressions on its upper surface, and a spiny margin. Powerful pincers are held close to the body, also equipped with a good 'cutting' edge. Carapace measures 6 cm across.

40 Fossil spider, *Abliguritor niger*, Oligocene. This spider is encased in amber. The colour of amber is related in part to its age – darker amber is usually older. The spider, less than 1 cm long, is perfectly preserved except for its soft parts. This fossil is known only from Baltic amber, the source of many famous fossil arthropods.

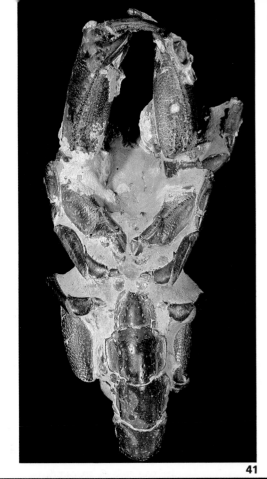

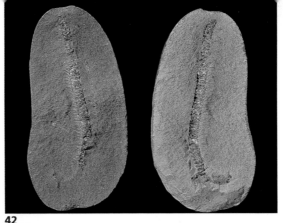

41   **43**

44

41 Fossil lobster, *Thalassina anomala*, Pleistocene, eastern Australia. The original shiny cuticle of the lobster is here preserved in a yellow, limy matrix. All the legs are attached, some partly concealed. Ths species has long thorny pincers with little upturned 'fingers' at the end. It also has the distinction of being the youngest fossil in this book. Length, about 12 cm.

42 Fossil millipede, *Euphoberia ferox*, Carboniferous. This fossil millipede is preserved inside a nodule of siltstone (part and counterpart of the specimen are shown). Each pair of legs, shows as white, hair-like lines on this specimen. This specimen is from the coalfield of Yorkshire, England. Similar species occur at Mazon Creek, Illinois, USA. Length 6 cm.

43 Spiny odontopleurid trilobite, *Acidaspis coronata*, Silurian, Worcestershire, England. This spiny little trilobite is preserved in a fine-grained hard limestone. Most of the original detail of the skeleton is still visible. Trilobites of *Acidaspis* type have thoracic segments extended into long spines, which also fringe the head and tail. Complete specimens are generally 2–3 cm long.

44 Fossil dragonfly, *Cymatophlebia longialata*, Jurassic, Solenhofen, Germany. The dragonfly is splendidly preserved on the flat bedding plane of a limestone. Much of the fine detail is preserved, especially on the wings. Dragonflies have stiff, long wings, with finely netted veining, long slender bodies, and large eyes, which are well seen on this fossil. Longest wing here has a length of 7 cm.

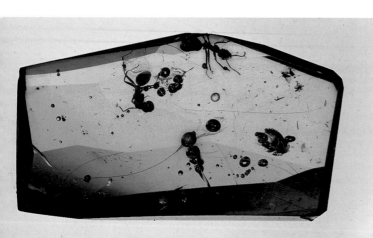

45 47

46

45 Fossil ant, *Iridomyrmex geinitzi*, Oligocene. Ants frequently got caught in the pine resins that were destined to become amber, and numerous fossil species have been recognized. This specimen is from Baltic amber. Living *Iridomyrmex* are found principally in Australia and South America, with a few species in India. Length about 4 mm.

46 Fossil lungfish, *Dipterus valenciennesi*, Devonian. The specimen is flattened, 18 cm long, but otherwise exceptionally complete on a bed of fine silt-stone. It has a long body with overlapping scales, with a concentration of fins at the hind end. This specimen is from the Achanarras fish bed, Caithness, Scotland, a famous locality.

47 The armoured fish, *Bothriolepis canadensis*, Devonian, Quebec. The trunk of this armoured fish is preserved in a fine-grained siltstone. The original bony material is well-preserved. The 5 cm long trunk is composed of a few, very large plates which have a tuberculate ornament. In front there is a pair of powerful, flattened appendages which are hinged on the side of the body

48 Eocene bony fish, *Pristigenys substriatus*, Monte Bolca, Verona, Italy. This exquisite fish is preserved flattened on the bedding plane of a flaggy lime-stone. The original outline of the fish is clearly visible. This fish has a deep body, narrowly com-pressed from side to side, very large eyes, prominent fins on both the upper and lower surfaces of the body, and a fan-like tail. From tail to mouth the fish measures 10 cm.

48

These other groups of reptiles lived alongside the more spectacular dinosaurs during the Mesozoic: turtles, lizards and crocodiles

The dinosaurs have dominated the popular conception of the fossil reptiles, sometimes to the extent of seeming almost synonymous with the word 'fossil'! The group includes by far the largest of terrestrial animals ever to have lived (*Brachiosaurus*), creatures to dwarf the largest bull elephants. The predators that preyed on the giants were even more spectacular, and by now the name of *Tyrannosaurus* is so well-known that it seems to be one of the first tongue-twisters mastered by small children. Of course there is little chance of an amateur collector acquiring an appreciable part of one of these great animals, but in the right site fragments of dinosaur skeletons may be picked up. The term 'dinosaur' includes a number of really rather separate groups of animals, with more or less independent evolutionary history during the Jurassic and Cretaceous. The real giants were the lumbering *sauropods* (*Diplodocus, Brachiosaurus,* etc.), four-legged dinosaurs, with enormously long necks only matched by equally long tails. The giant predators that walked on their hind legs (*Tyrannosaurus, Allosaurus* and other theropods) share structures of the hip bones with the sauropods that show they are more closely related to them than to the rest of the dinosaurs. These latter (Ornithischians) include some animals that walked on their hind legs like *Tyrannosaurus*, but with vegetarian habits (*Iguanodon*, the first dinosaur to be discovered); the group also includes various types of vegetarian armoured and plated dinosaurs, often shown in pitched battle with their carnivorous contemporaries. These well-protected dinosaurs each diversified into a number of genera; the three main groups are: the stegosaurs of the Jurassic, with paired plates along the back and a nastily spiked tail (*Stegosaurus*); the ceratopsians, horned, rhinoceros-like dinosaurs of the Cretaceous, including the familiar *Triceratops*; the spiky armoured ankylosaurs, tanks on stumpy legs. Specialized ornithischians in the late Cretaceous were the remarkable duck-billed dinosaurs, animals that lost their front teeth and had arrays of tiny grinding teeth at the back of the jaw that were continually replaced, like those of the shark, and must have been able to cope with tough vegetation. Some of the duck-billed forms evolved crests and protuberances on top of their heads. All the dinosaurs, vegetarian and carnivore alike, became extinct at the end of the Cretaceous period (*see* p. 135).

A whole range of other reptiles were present in the Jurassic and Cretaceous; none of them are dinosaurs. They include the products of the spread of the reptiles back into the aqueous environment from which their distant ancestors emerged, and, for the first time in the history of the vertebrates, the conquest of the air. The marine *plesiosaurs*, with their long flexible necks and sharp teeth, were formidable hunters of Mesozoic fish. The limbs were modified into efficient paddles, perfectly adapted for sculling through the water. Even more streamlined for marine life were the ichthyosaurs ('fish lizards'), which, as their name implies, include species that look remarkably fish-like, although perhaps the better analogy would be with the porpoises, a group of mammals that 'returned to the sea', and may fill a similar role in modern seas to that of the ichthyosaurs in the Jurassic. These marine reptiles have a higher chance of preservation than many of the terrestrial species, and their carcasses often seem to have sunk to the sea floor, so that some localities have yielded numerous complete specimens. The Liassic (Lower Jurassic) rocks of Europe furnished many of the magnificent specimens which are now on display in museums. Some of the ichthyosaurs include baby individuals within their skeletons – which either implies cannabalistic habits, or, more probably, the ability to give birth to their young alive. Isolated vertebrae of marine reptiles (*see* Fig. 50) are one of the reptilian fossils that are the easiest to collect.

The pterosaurs, flying reptiles, also had their origins in the Triassic and became widespread and varied during the Jurassic and Cretaceous. Like the bats, the pterosaurs had membranous wings stretched between highly modified arms and 'hands' and the fore part of the legs; unlike the bats some of them grew to an enormous size and the front support for the wing was provided by only one finger of the hand, grotesquely extended. Most of them, and particularly the larger ones, combined slow, flapping flight with gliding, spending a large part of their lives effortlessly aloft. As in the case of the sauropod dinosaurs they seem to have got bigger and bigger during the Cretaceous – some of these later pterosaurs are supposed to have had wing spans exceeding 10 metres, which would make them almost comparable

to a man-made glider. Why this large size evolved is something of a mystery: it may have been that larger gliders cruise at slower speeds without 'stalling'. Quite how these amazing animals landed and took off (if they did) can only be guessed at. Like the plesiosaurs and ichthyosaurs, they were successful and varied for a long time, but all three groups failed to survive the Cretaceous.

Other groups of reptiles were not exterminated at this time, even though their fossils may be found in rocks as old as those that yield dinosaurs and the other spectacular, extinct groups. Crocodiles and lizards have a fossil record extending back to the Triassic, snakes to the Cretaceous. The turtles and tortoises have almost as long a history as the crocodiles. The fused, bony plates that protect their soft parts make

Examples of the two major kinds of dinosaurs, showing the structure of the hip bones which distinguishes them. *Above,* saurischian (*Tyrannosaurus*), *below,* ornithischian (*Scelidosaurus*)

them well-nigh invulnerable. All the long survivors show every sign of flourishing today: one might suppose that the tortoise will continue its lumbering progress no matter what catastrophes may come to pass in the future.

MAMMALS – CLASS MAMMALIA

Mammals – animals with mammary glands that suckle their young – were certainly present during the Triassic, and were a continuous if subordinate component of fossil faunas throughout the acme of the dinosaurs. It was only after the demise of the ruling reptiles that the class had the opportunity to capitalize on the evolutionary advantages which propelled them to the dominance they have enjoyed ever since. Many of the early mammals were probably insect eaters, as are many of the primitive representatives of the group today. Small mammals were present in the Triassic – shrew-like animals with long noses, and probably insatiable appetites for smaller items of the fauna. We can imagine animals like these darting through the undergrowth in search of food while the colossal reptiles lumbered obliviously around them. It would be wrong to suppose that there was no evolutionary activity among these early mammals: in fact there were considerable changes, which by the late Cretaceous had established early members of several of the principal groups of mammals which dominate the terrestrial world today. The marsupial (pouched) and placental (womb-bearing) mammals had separated, and lived side by side.

Many of the early mammalian remains are fossil teeth. More than those of any other vertebrate group, the teeth of mammals have changed in distinctive ways, and the different types of mammals can be reliably identified from a study of their dentition alone. This is particularly fortunate, because teeth have a high fossilization potential, exceeding that of other parts of the skeleton. The sediments filling caves or fissures in limestone are frequently repositories for fossil teeth. Sampling for mammal fossils often involves the patient sieving of great quantities of sediment to extract the fossil teeth. If you are very lucky you might find a whole jaw (*see* p. 100). Mammal jaws often have specialized teeth for particular jobs – nipping, chewing, biting, gnawing teeth and so on – which function in perfect cooperation between the upper and lower jaws. Mammal teeth often have distinctive patterns of crests or bumps (*cusps*) that serve to identify large groups within the mammals. Teeth for grinding vegetation are often large and ridged, while the rodents have their fast-growing gnawing teeth at the front of the jaw. The most specialized dental structures of all are probably the baleen plates of the largest whales, modified for straining plankton from sea water.

The evolution of mammals after the Cretaceous was both extremely rapid and very complex, and we can do no more than give the roughest sketch here. Compared with what was happening in contemporary marine invertebrates, the visible changes in mammalian skeletons during the early part of the Tertiary were fast and dramatic; it did not take long before the roles of larger herbivores and carnivores were filled by mammalian animals. By the end of the Palaeocene, only a few million years from the Cretaceous extinction of the giant reptiles, there were representatives of many of the living mammalian orders, including, for example, the primates (the order to which man belongs), the carnivores (cats dogs, and most living predators) and the rodents (rats, mice). The history of the mammals in the Tertiary involved not only this one major diversification, but several, with major extinctions between, so that the largest mammals of the earlier parts of the Tertiary are not related to the large mammals of the Recent. The roots of our present European fauna extend back to the Miocene. The whole story is complicated by the fact that this was also the time of break-up of the supercontinent Pangaea. Mammalian faunas on isolated continental fragments could evolve, at least for a time, separately from faunas in other parts of the world, producing a whole series of peculiar animals that have no direct relationships to animals with a similar mode of life elsewhere. The obvious example is Australia, which was separated early, and in which the marsupial mammals had the opportunity to adapt to a whole range of ecological niches, which they managed with remarkable success in spite of a low cranial capacity and a primitive mode of reproduction. Some of the largest marsupials (*Diprotodon*) seem to have survived to a time tantalisingly close to the present. Similarly, in South America a whole range of peculiar mammals evolved largely in isolation, including giant sloths, specialized grazing species, and the extraordinary glyptodonts (*see* p. 100). A lot of argument among mammalian palaeontolo-

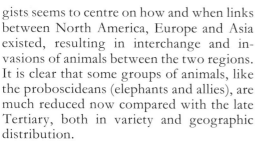

the widespread cold conditions, roamed widely around the northern hemisphere, including woolly mammoths and rhinoceros, and extinct species of tigers. On the other hand interglacial periods were very much *warmer* than conditions in temperate latitudes today, and these climatic fluctuations resulted in periods when hippopotamus and tropical rhinoceros thrived in the areas occupied by Europe's capital cities today. With the final retreat of the ice many of the cold climate specialists perished (possibly with the help of man), although some of the species that accompany them, like lemmings or musk ox, survive today in the harsh conditions of the arctic tundra.

BIRDS – CLASS AVES

Fossil birds are distinctly rare, and to find a well-preserved and complete example is a palaeontological event. The evolution of flight led to generally light and fragile skeletons, and it is perhaps not surprising that avian fossils are so uncommon, and when found, so often fragmentary. The derivation of the birds from a reptilian (possibly dinosaur) ancestor is certain, and because the earliest fossil bird, *Archaeopteryx*, is Jurassic in age, it is usually assumed that the split from the reptiles occurred in the earlier part of the period. Unfortunately, there is much that is not understood about the history of the birds between *Archaeopteryx* (*see* p. 101) and the good assemblages of fossil birds undoubtedly related to living species in the

gists seems to centre on how and when links between North America, Europe and Asia existed, resulting in interchange and invasions of animals between the two regions. It is clear that some groups of animals, like the proboscideans (elephants and allies), are much reduced now compared with the late Tertiary, both in variety and geographic distribution.

A late stimulus for the evolution of distinctive mammals was the Pleistocene Ice Age, although a variety also became extinct at this time. A number of species, adapted to

Right The Berlin skeleton of the early bird, *Archaeopteryx*

Left Some of the specialized mammals which evolved in isolation in South America: top *Glyptodon*; middle, *Toxodon*; bottom, *Megatherium*

Below Lower jaw of Pleistocene bear showing typical mammalian specialization of teeth: sharp canines, chewing teeth at the rear

Eocene and Miocene. Cretaceous birds are generally scarce, and some of the 'toothed' birds that are known, like the flightless, aquatic *Hesperornis*, seem to be difficult to relate to any of the living species. It is probable that like the mammals the roots of the great variety of living birds are to be found in the Cretaceous, but fossils, which could document this, are only now slowly coming to light.

FOSSIL PLANTS

The fossils of plants are some of the most attractive that the amateur collector is likely to find. Those plants that concern us here were terrestrial or lived in bogs, rivers or lakes. Like animals that dwelt on land, the terrestrial flora also arose from ancestors that lived in the seas, a change that is known to have happened before the end of the Silurian. However, marine algae were very important in the early history of the evolution of the earth, and they are described in some detail in Chapters 7 and 8. They are not generally conspicuous fossils in the field. A few algae, however, produced hard calcareous skeletons, and were abundant enough to produce small gardens beneath the sea. Such *calcareous algae* (*see* Plate 52) have a long history, extending back into the Precambrian, and they are still numerous today. They were often components of reefs, and in some environments it is the algae that bear the full brunt of the attack by the breakers on the exposed, seaward side of the reef. They are also important rock builders; in northerly latitudes today limestones are being formed from the skeletal debris of such calcareous algae as *Lithothamnion*.

The first land plants must have been derived from a marine alga. We have to visualize this as a very gradual process, the early transitional plants perhaps creeping across mud flats, or partially submerged. The change involved the production of a protective 'skin' to ensure that water loss to the atmosphere was not excessive, and the stems had to acquire sufficient rigidity to stand up without the support of surrounding water, but also the cellular structure had to allow the passage of nutrients to the growing shoots (ultimately, a *vascular* system). The evolution of such a plant could not have been achieved at a single stroke. Reproduction in these early plants was by means of minute spores. Quite a variety of Silurian plants are now known, and by the end of the Devonian it is apparent that most of the problems of terrestrial living had been solved, to the extent that large tree 'ferns' of the time would have had dimensions comparable with forest trees today.

The fossil record of plants is not good enough simply to 'read out' the story of their evolution by collecting from successively younger, non-marine rock formations. Some of the earliest plants do not appear to be the most primitive, and *vice versa*. True, there is a general drift through time towards more complex plants, more perfectly adapted to life away from water, but primitive species have persisted alongside the innovations. The fossil record is full of curious plants that do not fit comfortably into classifications – this is one of the things that makes the study of fossil plants so fascinating. Consequently there are arguments among botanists about the relationships of many plants, living and fossil, and some classifications have quite large categories containing only a few plants, which may be rare. Not all of the major groups will be described here, but only those that are likely to be found by the non-specialist collector. Some important groups, like the mosses and liverworts, have a sporadic fossil record extending back to the Carboniferous, but they are not likely to be encountered without a special search.

Plant fossils are often to be found in particular beds, reflecting conditions of deposition that were just right for their preservation. Of course, the rocks that contain abundant plant remains are mostly those which were deposited in freshwater lakes, or on deltas, or the fossilized remains of peat beds. There were probably plants living in hilly regions from quite early which have left little fossil record, and this may account for some of the gaps in the preserved fossils. Many fossil plants are now more or less black, carbonized ('coalified') compressions. In some cases, however, the fine structure, even down to individual cells, is splendidly preserved, and these are the most important specimens from a scientific point of view. The problems of associating the different organs of the same fossil plant species were mentioned earlier in this book (p. 23).

EARLY LAND PLANTS – DIVISION PSILOPHYTA

The earliest land plants, to be found in Silurian and Devonian rocks, had simple shoots that arose from a creeping 'axis', which were little different in structure from the shoots themselves. Leaves, if they were

Right Seed fern, *Trigonocarpus parkinsoni*, Carboniferous. These fossil seeds are preserved in a relatively coarse-grained sandstone. Although they are uncrushed, only parts of the original seed covering remain – so that they are largely internal moulds. They are the seeds produced by the fern-like plants such as *Mariopteris*, but because they are difficult to link with their foliage they have been given separate names. The three raised ridges on their surface are typical. Seeds of similar type have been found associated with a number of the Carboniferous seed-fern leaves. They are from Ayrshire, Scotland; length, 2 cm.

spores, which form in clusters on the undersides of the fronds, or sometimes on specially modified fronds. To be really sure that one is dealing with a fossil fern it should be possible to see the spore cases, because other kinds of plants can produce fern-like foliage. In practice, many of the fossil fronds that one finds are sterile, and for these there a number of *form genera*, units of classification based on the shape of the fronds alone (*see* p. 23).

Although many ferns are low herbaceous plants, they have periodically attained the dimensions of trees; forests of such tree ferns exist today in humid regions, for example in New Zealand. Large ferns were present as early as the Upper Devonian and different genera as large or larger were significant components of forests during the Carboniferous to Cenozoic. In these cases one never finds a whole tree fossilized, and it is necessary to piece together the whole plant by making intelligent guesses about whether a frond of one kind is persistently found with a particular fossil trunk.

SEED FERNS – DIVISION PTERIDOSPERMOPHYTA (Colour plate 57)

Many of the fern-like fronds that can be found in Carboniferous rocks belong to an extinct group of plants sharing a common ancestor with the ferns, but distinct enough to be regarded as entirely separate. These seed ferns had much-enlarged female reproductive bodies – forming true seeds – which were fertilized by pollen born on separate organs. Some of the seeds hung like bells from the ends of the ferny frond, and in some species they were several centimetres long. The plants were generally bushy shrubs, but some may have been climbers, and others robust enough to form small trees. A lot of the species belonging to the commonest of the Carboniferous and

present at all, were humble scales or spine-like projections. The upright twigs often branched in a simple way, forking into two, and then into two again, and sometimes terminating in little capsules that carried the spores. The plants are often preserved as dark markings on shales, and were less than half a metre in length. The best-preserved examples, from which their cellular construction has been described, are from the Devonian Rhynie Chert (p. 17).

FERNS – DIVISION PTEROPHYTA (Colour plate 60)

The feathery and delicate fronds of the ferns make them among the most beautiful of the foliage plants, and they attract a fanatical group of devotees dedicated to growing exotic species in inhospitable cities. They are also a very ancient group, with many Carboniferous representatives, and a number of 'preferns' in the Devonian as well (Plate 59). Then as now they generally preferred humid environments in which to live. They reproduce by means of minute

Right Some of the commoner types of fern-like foliage from the Carboniferous. Such foliage often belongs to the seed ferns, rather than the true ferns.

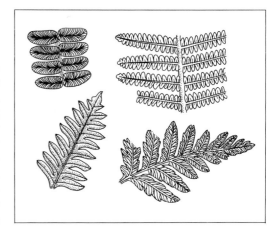

Fossil fern, *Onychiopsis mantelli*, Cretaceous. This delicate fern is preserved in a very fine-grained sandstone, which fractures rather irregularly. Slender and delicate branching patterns with many fine subdivisions are characteristic of this kind of fern. There are a few similar species, known rarely from rocks of Cretaceous and Tertiary age, distinguished on the branching patterns of the fronds. This specimen is from Sussex, England. Related forms occur in North America. The total length of the frond is 12 cm, but it was originally part of a much larger plant

Permian 'leaf genera' – like *Neuropteris*, *Alethopteris* and *Mariopteris* – were seed ferns of this kind (rather than true ferns), but associations of seeds with foliage are extremely rare. The seeds themselves (*see* p. 103) can be locally common. Although they are commonest in rocks of the Late Palaeozoic, some Mesozoic plants have been regarded as a later branch of the seed ferns.

LYCOPODS – DIVISION LYCOPHYTA
(Colour plates 53, 54, 56)

Many of the great trees of the Carboniferous were lycopods, the trunks of which contributed largely to the formation of coal. This large and important group of plants is represented today by a mere four genera of humble herbaceous plants – *Lycopodium*, *Selaginella*, *Isoetes* and *Phylloglossum* – living fossils which have survived with little change from the Mesozoic or even the Carboniferous. The lycopods have long, narrow leaves like little straps, each of which leaves a characteristic scar on the stem at its point of attachment. The plants reproduced by means of spores, which were born in cones; the spores were often of two kinds – large (female) megaspores up to 2 mm in diameter and much smaller (male) microspores. The study of the spores of these plants is important in correlating rocks of late Palaeozoic age. The large trees had extensive roots, which grew nearly horizontally through the soil, repeatedly forking, and often bearing long rootlets. Fossil roots (Fig. 56) are quite common, and may be preserved even when the trunks are not. The distinctive patterns formed by leaf scars are visible on fossils of the bark, which is often found in coal and in the deposits overlying ones containing roots.

Lepidodendron has many spiral rows of lozenge-shaped leaf scars, while *Sigillaria* has long rows of elliptical scars; both are very common fossils in association with coal measures. Cones are much rarer; these are about the size and shape of an asparagus spear, with rows of cavities, which contained spores, so that if they are flattened they can appear almost segmented in the manner of an arthropod fossil. Some of the best specimens of trunks were preserved cell by cell in silica, and these show even the finest details if they are cut and polished. The lycopods can be found in Devonian or even Silurian rocks, but they are generally uncommon in rocks younger than Permian, and all the tree species apparently disappeared before the end of the Palaeozoic.

HORSETAILS – DIVISON ARTHROPHYTA
(Colour plate 58)

On any walk through a bog or fen one can see the delicate shoots of the horsetails, looking almost like miniature fir trees, sometimes clustering into dense masses. If one plucks a piece from one of these plants it will almost certainly break along one of the numerous joints that divide the whole plant into segments, right to the tips of the branchlets. These joints give the arthrophytes ('jointed plants') their scientific name, and the little whorls of tiny branches produce the bushy appearance which accounts for their popular one. There is only one living genus of horsetails – *Equisetum* – although its many species are very widespread. The group were formerly varied and numerous, and even included trees perhaps 20 metres in height. The horsetails reproduce by means of spores, which are born in small cone-like structures at the tip of the branches or on modified shoots. They are among the easiest of fossil plants to recognize, for even flattened stems retain the characteristic jointing, and the feathery distal branches cannot be mistaken for anything else. The flattened whorls are particularly easy to recognize, and are frequently encountered in the soft, dark shales associated with coal seams. As with so many of the primitive plants the arthrophytes had their heyday in the Carboniferous after a probable origin in the Devonian. Because they are common in aquatic environments they have a rather good fossil record, and are not uncommon in freshwater sediments of Jurassic and Cretaceous age. It is a mystery why only one genus of this formerly diverse group should have had the capacity to survive to the present day.

CYCAD-LIKE PLANTS

In the forests of the Mesozoic there were large numbers of trees and shrubby plants that looked at first glance like palm trees, with leafless stems crowned with bouquets of stiff, large leaves, deeply divided into long, unbranched leaflets. These were cycads and bennettites. A small number of cycad genera survive in the tropical and subtropical regions today, but they are much less conspicuous today than they were 120 million years ago. The resemblance to palms is no indication of their true relationships, because the cycad-like plants were not true flowering plants like palms. They do, however, have a most extraordinary fructification, a large knob arising from

Reconstruction of a Carboniferous coal swamp, showing large lycopod and cordaite trees, an amphibian, giant dragonfly, and millipede

the centre of the crown, looking something like a corncob, and bearing numerous, very large seeds. Nowadays, botanists divide the cycad-like plants into two major groups which may not be particularly closely related: the Cycadales, which include the living species, and the Bennettitales, an important Mesozoic group of plants which have smaller fructifications scattered among the bases of the leaves. Marie Stopes, who later became a famous pioneer of birth control, started life as a palaeo-botanist interested in the cycads. Some specimens of the trunks of bennettites (*see* p. 107) are among the most spectacular of fossils, and they even attracted the attention of the Etruscans, who included them among their sacred relics more than four thousand years ago! The amateur collector is most likely to come across the cycad foliage, with its characteristic narrow central rib and flat leaflets. Their fossils are commonest in Triassic, Jurassic and Cretaceous rocks.

CONIFERS – DIVISION CONIFEROPHYTA
The conifers are as varied and successful a group of plants today as they have ever

been. But again they are an ancient group with ancestors back in the Carboniferous forest, a melting pot for plant evolution. Botanists are not fully in accord about how many separate groups of plants there are within the conifers, and they are all lumped together here. The fir cones of modern species will be familiar to the reader, often littering forests in their thousands, and containing seeds, which may be equipped with wings. The close, needle foliage of most modern conifers, which seems as much immune to desert heat as arctic cold, was probably derived from lusher leaves in the later Palaeozoic. In the Carboniferous swamps large trees with leaves, of the genus *Cordaites* grew to 30 or more metres in height, with long slim trunks terminating in branches that bore strap-shaped leaves up to a metre in length. The cordaite 'cone' was a longer and more complex structure than that of any living conifer. The leaves of this tree are rather common fossils, striped with longitudinal veins that give it a superficial resemblance to the leaf of an iris. Most of the steps leading to the modern conifers

Large trunk of a bennettite *Cycadeoidea*, Cretaceous, USA. These conspicuous fossils are often over half a metre across.

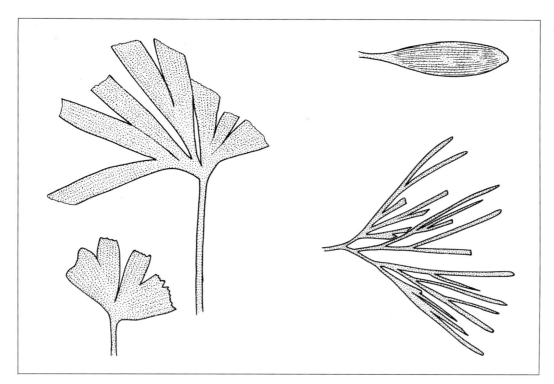

Left Varied shapes of leaves of fossil ginkgos, typically Mesozoic, with photograph of fossil example, natural size (*below*).

Right Frond of a cycad, *Nilssonia kendalli*, Jurassic. The fossil is preserved on the surface of a fine-grained siltstone. There are many other fragments of plants in the same block, but the frond is unusually complete. Cycad leaves have characteristic flattened leaflets arranged in rows on a very slender axis. There are a number of species of *Nilssonia* from the Jurassic distinguished on the form of the leaflets. Related genera include many species with a wide distribution, and there are living representatives occurring in tropical regions today. This specimen is from the famous plant beds of Yorkshire, England. Other well-known bennettite localities are the Black Hills of Dakota and from Maryland, USA. The frond has a length of 15 cm.

seem to have taken place by the end of the Palaeozoic; no other group of organisms of such antiquity has retained such an unbroken hold over vast areas of the earth as the conifers. Some primitive conifers, like the monkey-puzzle tree *Araucaria*, have survived with little change since the Triassic.

The long-leafed plant *Glossopteris* (Plate 55) is something of a botanical puzzle, but probably belongs within this broad group. It is common in later Palaeozoic rocks of South Africa, peninsular India, Australia and South America, and is thought to have lived almost exclusively in the cooler areas of the supercontinent Pangaea (p. 32).

The ginkgos are often placed in a separate group. They are represented today only by the maidenhair tree *Ginkgo biloba*, one of the obligatory 'living fossils' to be found in every botanical garden, and many parks in inner cities. Ginkgo leaves are particularly beautiful, perfect fans with splayed veins, and specimens of this general type are found in many localities from rocks of Triassic age onwards (*see* p. 108). The Mesozoic ginkgos were a varied and numerous group of plants, but unlike the spruces and pines they slowly declined in variety throughout the Tertiary.

FLOWERING PLANTS – DIVISION ANTHOPHYTA (Colour plates 61–64)

The flowering plants dominate the floral world today, except perhaps in the lonely spruce stands of the Arctic. They are the broad-leaved trees, the grasses, and myriads of tiny creeping plants, as well as the showy flowers which attract man and pollinating insects (p. 91) alike. Their diversity is truly extraordinary: in tropical forests there are hundreds of species of trees alone. Yet a good deal of this diversification probably happened in the last 70 million years, during the Tertiary. However, like the mammals and the birds, the flowering plants certainly originated, and probably did much of their evolutionary groundwork, during the Mesozoic and especially the Cretaceous. The details of their origins are still not clear. For one thing, flowers do not readily fossilize – and leaves alone can be misleading. The flowering plants (angiosperms) appear almost ready-made in the Cretaceous, and many of these fossils can even be placed into living genera. There is still much to discover of their even earlier history, and we hope that the crucial steps did not all take place in some site where fossils have little chance of preservation.

Whatever their early history, fossils of flowering plants – leaves, seeds and wood – are common in rocks of Tertiary age. They do not even have to be rocks deposited under fresh water conditions, because wood and seeds are perfectly capable of drifting long distances before becoming waterlogged enough to sink to the bottom of the sea. Leaves, however, tend to be preserved in the deposits laid down in lakes and the like. Where the sediments are fine-grained the leaves form fossils of exquisite beauty,

Oligocene sumac tree, *Rhus stellariaefolia*. These leaves are beautifully preserved in a flat-bedded and very highly fissile shale, laid down at the bottom of a fresh water lake. Long, graceful leaves are divided into about twelve spindle-shaped leaflets with undivided margins, and a terminal leaflet. Each leaflet carries a prominent vein in the middle. A few other fossil species of *Rhus* are known from the Tertiary rocks. Living sumacs are familiar in temperate regions, their leaves turning a beautiful dark red late in the year. The specimen shown here is from the renowned Florissant beds of Colorado, USA. Other fossil *Rhus* have been recovered from Eocene rocks in Colorado, Utah and California. The longest leaf here is 12 cm long.

preserving even the finest veins, which can be directly compared with the leaves of their living relatives. Another mode of preservation of Tertiary plants is beneath flows of lava (Oregon and the island of Skye, for example) where whole floras can be preserved in enough detail to reconstruct the detailed botanical ecology of the time. Fossil plants are of great use in plotting the many shifts of climate that happened during the Tertiary: a humid-tropical flora may be found in what is now an arid region, or a warm-climate flora may be found in an area that is now decidedly cool. This makes the assumption, of course, that the plants have not changed their habits since they were fossilized. During the dramatic climatic fluctuations of the last ice age – warm to cold to warm repeated several times – the flowering plants acted as thermometers for the climate, sensitive recorders of the shifts that affected everything from beetles to man.

CHAPTER FIVE

BRINGING FOSSILS BACK TO LIFE

In this chapter some of the ways in which palaeontologists determine the way fossil animals lived are described, reanimating the dead fragments to build up a living creature. Since popular ideas of life in the past are often founded on vividly coloured reconstructions of 'The World in the Jurassic Period' and the like, it is important to remember that these imaginative scenes are all inferences from bones, and similar fragments. In fact there is nothing fixed about such interpretations, because the way the fossils are understood may change over the years, but it is usually some time before new discoveries percolate into the popular presentations. Even the most solid-looking dinosaurs may have changed their habits in the last few years!

Many different lines of evidence may be used to flesh out the bare bones of the fossils. The first of these is the evidence of the rocks from which the remains were recovered. The sediments themselves reveal much about the environment of deposition, as was shown in Chapter 3. It is important to establish whether the fossil animal actually lived in the environment which furnished its sedimentary cover, or whether its remains were swept in from some other place. Fortunately it is usually easy to spot such intruders. The type of sediment and the associated fossils will show whether the environment was marine, freshwater or terrestrial, providing the basic information into which the ecology of the animal has to be accommodated. The sediments themselves may have preserved some of the tracks left by the animal (p. 20) to give direct evidence of its past activities. From the tracks alone it is possible to be certain of the bipedal stance of certain dinosaurs, and to measure their stride.

The character of the rocks, and their setting in the past geography at the time when they accumulated, diagnoses the climatic setting in which the extinct fauna lived, and climate imposes certain restraints on possible modes of life. Savannah animals differ from those of tropical rain forests, and these again from inhabitants of the tundra at high latitudes.

Looking at the fossil animal itself, the first necessity is to reconstruct it as accurately as possible from its fragmentary remains. Sometimes this is a very complex business, particularly for vertebrates with numbers of small bones. To proceed from the reconstruction to an assessment of probable life habits two different but complementary approaches are used. One method attempts to compare the structures of teeth or limbs or some other feature of the extinct animal with living *analogues*. These do not have to be biologically related organisms. The basic argument is that structures that are closely similar were probably adapted to a similar function. Sometimes these similarities are quite obvious: the ferocious teeth of a predatory dinosaur are a sure indication of hunting habits, with hardly a glance at the fangs of living mammalian carnivores. But the structure and functioning of grinding or chewing teeth in other mammals, in which opposing teeth co-operate in action, and which can be matched in extinct, unrelated mammals, is a much more subtle matter, involving detailed studies on the operation of living dental systems to help elucidate the functioning of fossil ones. The technique of 'hunt the analogue' is a favourite one practised by palaeontologists, but it is certainly not a foolproof one, because there are many fossil animals that defy comparison with living organisms, and some analogues do not stand up to detailed scrutiny. The second method tries to analyse the structure of the fossil almost as if it were a piece of engineering. If the fossil is constructed in a certain way, then there are only a limited number of 'jobs' that the structure could perform. The idea here is to decide which of the possibilities is the most likely one. This assumes that nature only manufactures efficient designs. For the most part this is a reasonable assumption to make. Most animals today do seem to have bodies that accord well with the functions they have to carry out to survive: flyers are aerodynamically efficient, active swimmers have suitable streamlining, herbivorous mammals have teeth appropriate for grinding plant food, and so on. Ideally one could construct a model of the fossil to test out these various functions in experiments, but the number of examples where the analysis has been pursued this far are limited. Of course it can never be known whether the right answer has been reached (unless somebody dredges up a 'living fossil' from the depths of the ocean) – there are only varying degrees of probability. But the best answers are probably obtained where the functional design of the fossil points to the same life habits as a living analogue, and where both are consistent with the geological circumstances in which the fossil remains occur.

SWIMMING TRILOBITES

There are thousands of different kinds of trilobites. All of them were marine and all of them are extinct. Since many kinds of trilobites co-existed at any one time they were probably occupying different ecological niches, behaving in different ways, corresponding with the wide variation in shape that they show. We can use the methods described above to elucidate some of these occupations, and get a glimpse, albeit an imperfect one, of the trilobite as it lived. I will take an example where *all* the methods can be used, and where all point in the same direction, so that the answer is probably correct. Most trilobites were broadly oval-shaped being rather longer than wide, not greatly convex, and with eyes that occupy perhaps a quarter of the length of the head. There are a few trilobite species, however, with enormous, globular eyes (*see* below). By examining the way these animals were put together it is possible to suggest a likely mode of life.

We can take the eyes first. The trilobite eye was of a compound type, and each lens was made up of a calcite crystal. From the optical properties of this mineral we know that the lens was able to interpret light coming from a direction more or less at right-angles to the lens surface. So we can deduce what the field of vision of our globular-eyed trilobite was, by looking at the directions in which all the lenses face. It turns out that our animal was able to see in almost every direction – upwards, downwards, sideways and forwards, and even backwards, because the eyes bulged out beyond the line of the rest of the body. Most other trilobites have a predominantly lateral field of view. If the animal lived on the sea bottom, it seems unlikely that it would have eye lenses specialized for looking *downwards*, and we begin to suspect that the animal habitually dwelt above the sea floor. Other features of its shape are consistent with this. At the

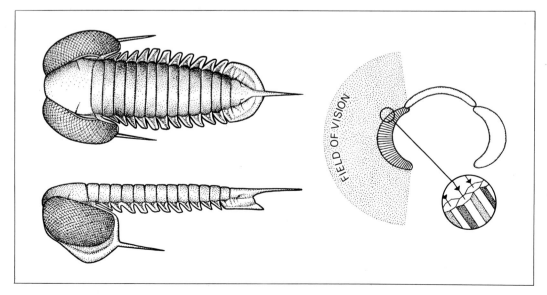

Giant-eyed Ordovician trilobite, seen from the top and side. Twice its natural size. Enlargement shows a section through the lenses to show the direction of the field of view.

FIELD OF VISION

49 Fossil bony fish, *Lates gracilis*, Eocene. This fish is preserved in a fine-grained limestone which preserved most of the skeletal details of the fish (but not its colours). This is a stout-bodied fish, something like a perch in general proportions, with long, low fins on upper and lower rear of the body (fin *rays* finely divide near their ends). Head to tail the fish measures 17 cm.

50 Vertebra of the Jurassic pliosaur, *Liopleurodon* species, Kimmeridge, southern England. This is only one vertebra of an animal much too bulky to appear in a book of this size! The original porous bone has been consolidated to become much heavier than it would have been in life. Diameter of the vertebra is about 10 cm.

51 Fossil frog, *Rana* species, Miocene, Spain. This species is preserved in a soft shale, but one which allows for the skeletal anatomy to be preserved in its entirety. The outline of the soft parts are also clearly displayed, so that the animal has been preserved almost in frozen motion. Frogs are one of the more numerous of amphibian fossils. Specimen 12 cm long.

52 Fossil calcareous alga, belonging to the genus *Coelosphaeridium*, Ordovician, Ringsaker, Norway. These spherical algae are preserved as internal moulds in a mudstone, the original calcite having disappeared. Canals within the algae stand out as rods in this kind of preservation, which is common in Ordovician rocks. There is a central sphere visible on one of the specimens. Related forms are spread over much of the world in Ordovician and Silurian times. Individual globes have a diameter of 1½–2 cm.

49

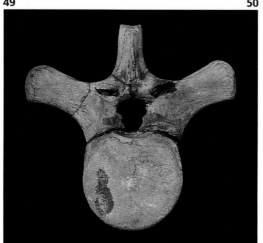

50

51

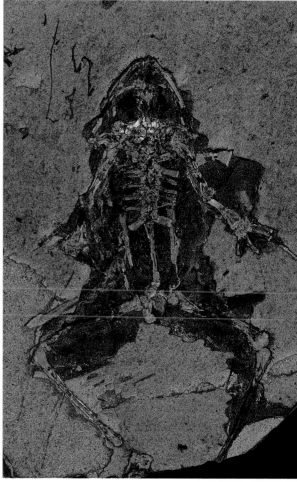

52

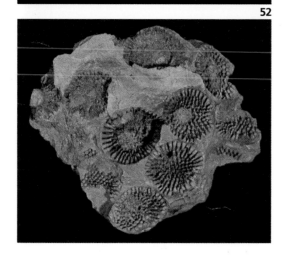

53 Fossil lycopod bark, *Lepidodendron rimosum*, Carboniferous, Newcastle, England. This is a typical 'coal shale'; shales and silts occurring between productive coal seams often contain impressions of the plants that went to form the coal. The plant material itself is converted to black carbon. This fossil bark is typified by its lozenge-shaped leaf scars arranged in spiral rows. This fragment is 10 cm long.

54 Fossil bark of the lycopod tree, *Sigillaria laevigata*, Carboniferous. This specimen lies on a slab of mudstone, the specimen itself having been converted into coal-like material — that is, into carbon. *Sigillaria* bark is covered with little horseshoe-shaped scars (*leaf cushions*). It is found in North America (Canada and the US), England, Belgium, France, Germany and Russia. This piece is 9 cm long.

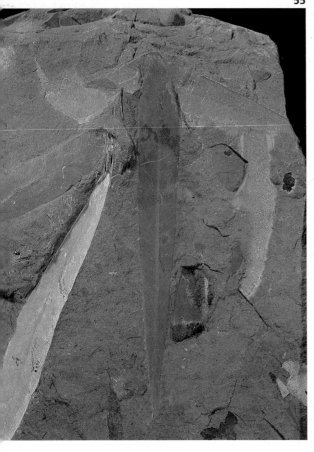

55 Leaves of the enigmatic plant, *Glossopteris linearis*, Permian. Many of these long leaves are preserved on the bedding planes of a siltstone. *Glossopteris* is characterized by its long strap-shaped leaves, which have a conspicuous midrib, with many fine, smaller veins branching off it to form a network. Its classification in relation to other plants is still unsolved. These leaves have a length of 7—8 cm.

56 Fossil root, *Stigmaria ficoides*, Carboniferous. This root is preserved as an internal mould, in a tough sandstone. It has been slightly crushed. The infilling sandstone is derived from the fossil soil over which the root was growing. *Stigmaria* is the name given to roots of the giant lycopod *Lepidodendron* (this page). This specimen is 15 cm long, but roots have been found which attain a length of 14 metres!

57 Leaf of a Carboniferous seed-fern, *Mariopteris muricata*, Staffordshire, England. The specimen is preserved in a hard claystone nodule, which preserved the leaflets in relief. This is quite a common preservation in coal measures. Microscopical details are not so well-preserved, and the fossil is, of course, only a fragment of a larger plant. The leaflets are attached to the axis along their entire base. 6 cm long.

58 Fossil horsetail, *Annularia sphenophylloides*, Carboniferous. This specimen is preserved as a carbonaceous compression on a very fine-grained and well-bedded sandstone, a preservation common in Carboniferous rocks. The little circlets of flat 'leaves' born at regular intervals on the jointed stems serve to distinguish this plant from others in the Carboniferous coal-shales. 8 cm long.

59 A 'prefern', *Archaeopteris hibernica*, Devonian, Kilkenny, Ireland. A fern-like plant, beautifully preserved in a yellow sandstone. The original material of the axis shows up black against the background of the rock. The large frond carries many branches, on which there are numbers of leaf-like *pinnules* — more or less alternating on either side of the branch. The specimen is 25 cm long.

60 Fossil fern, *Sphenopteris laurenti*, Carboniferous, Derbyshire, England. The frond of this fossil fern is preserved as a compression on the surface of a fine-grained, mica-rich siltstone. Some of the original fine detail on the leaflets is preserved. Reproductive structures should really be found to distinguish it from other, superficially fern-like plants. This frond is 9 cm long.

57 58

59

60

61   63

61 Fossil poplar leaf, *Populus latior*, Miocene. The leaf is excellently preserved on the flat bedding-plane of a limestone laid down under fresh water. Poplar leaves have an elegant outline resembling that of an arab minaret. This species has a finely toothed margin; a large, wide leaf born on a long stem. This type of tree has a history extending back to the upper Cretaceous. Length of leaf is 18 cm.

62 Miocene maple seeds, *Acer trilobatum*. Seeds of the maple family are typically in pairs, joined at the base, with flat 'wings' that help the seeds to disperse. Seeds of *Acer* are rare compared with the leaves, which can be found in rocks dating back to the Cretaceous period. This is from the Oeningen, Switzerland. Seeds are about 3 cm long.

62   **64**

63 Miocene maple leaf, *Acer trilobatum*. This leaf is exquisitely preserved in a fine-grained, flat-bedded marly limestone. Although much of the fine detail can be seen most of the tissue of the leaf is destroyed. The leaf is divided into three parts with a broad central lobe (finely toothed) and two flanking lobes. Simple, distinct veining can also be seen. The leaf is 10 cm long.

64 Fossil flower, *Porana oeningensis*, Miocene. These beautifully preserved flowers are from the Oeningen deposits like the maple leaf and seeds illustrated here. Fossil flowers are extremely rare because of their delicacy and because they are shortlived. These flowers have five oval petals, finely veined. No details of the stamens are present in the fossil state. Diameter of the flowers is about 2 cm.

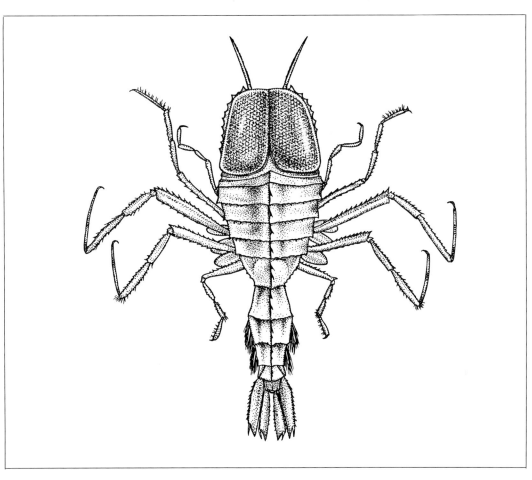

Giant-eyed living crustacean (*Cystosoma*)

edge of the eyes are a pair of long spines, and these point downwards at a steep angle, at a very awkward attitude for resting comfortably on the sea floor. Most 'normal' trilobites have a more or less horizontal rim around the forward margin, which may have rested on the sediment surface. The thorax of the large-eyed trilobite is remarkably long compared with the average trilobite (compare Fig. 33). The convex, middle part of the thorax contained the musculature that operated the appendages (which, as usual, are not preserved), and the relatively large volume of this region shows that the musculature was powerful. The construction of this giant-eyed trilobite suggests that it lived an actively swimming life well above the bottom, and possibly in the surface waters of the sea. Of course because there are no living trilobites we cannot find direct confirmation of this hypothesis, but we *can* look for other arthropods that have the same modifications of the eyes, for example, in the present day oceans. The analogy is found in such ocean-going amphipods (a kind of 'shrimp') as *Cystosoma*, which also has enormously expanded eyes, looking like headlamps, compared with its

bottom-dwelling relatives (*see* above). In fact many of the arthropods that inhabit the water column have large, globular eyes of this kind.

If it is correct that our trilobite lived above the sea bottom, actively swimming in the water, there are certain predictions we can make about its geological occurrence which can be tested by looking in the rocks. Most bottom-dwelling trilobites preferred to live at a particular water depth, or on a particular type of sea bottom (mud, sand or lime). No such restriction should apply to our large-eyed species; it should be found with all other different kinds of trilobite assemblages without preference. This has been proved in several places: in some arctic localities it is found with trilobites that lived at great depths in the muds of the Ordovician ocean, while in Canada the same species occurs mixed with the inhabitants of the shallow-water seas, where limestones were accumulating. By the same token the free-swimming trilobite may be expected to have a very wide geographical distribution, for oceans would be no barrier to it. And again this proves to be the case: our example is found in Arctic Canada, in the high Arctic island of Spitsbergen, thousands of miles away in what is now the desert country of Nevada and Utah in the United States, in western Ireland, in Russia, and in north-western Australia.

So in this case the different sources of evidence all point in the same direction, whether derived from a detailed consideration of the way the trilobite is constructed, or by analogy with living animals with similar adaptations, or from the evidence of the rocks or the distribution of the fossils. These giant-eyed trilobites of the Ordovician were active swimmers in the surface waters, as near certainly as we can ever know.

There are all kinds of different questions we can ask about the life habits of these trilobites, which are not subject to the same kind of careful scrutiny. What did they eat, for example? Because we do not have the legs preserved, it is not possible to see how the mouth appendages functioned in feeding, and neither are the stomach contents preserved. So here there are only more or less likely speculations. Active swimmers in the surface waters of the present oceans are likely to feed directly on the plankton, and the trilobites may have had a method of harvesting large quantities of minute food. Or they may have been hunters after larger prey, in which case when appendages are eventually discovered they may prove to have adaptations for grasping and manipulating larger food. Puzzles remain, even though we can be certain of the rudiments of the story.

CONTROVERSY AMONG THE GIANT DINOSAURS

The giant sauropod dinosaurs like *Brachiosaurus* were the largest land animals the world has seen – they weighed more than 80 tonnes – and present quite different problems in the interpretation of their life habits from the diminutive trilobites. Such spectacular animals have obviously attracted much attention, and one might suppose that the problem of how they lived would have been satisfactorily solved long ago. Most of the popular books showing the Mesozoic giants in their natural setting portray them wallowing about in swamps flanked by deep vegetation, their bodies largely under water. Surely an animal of this bulk, it was argued, must be partly supported by water, and their relatively inconsequential teeth must have been adapted for chewing on the kind of soft, luxuriant vegetation that flourishes in and around swamps. Because their nasal openings were on the top of their heads they could even continue to breathe if it became necessary (*see* p. 115) to totally submerge. Recently the life habits of the giants have been looked at in a way that disproves most of these traditional notions. Consider the structure of their legs. The sauropods have relatively long, pillar-like legs, resembling those of the elephant, the largest *living* land animal, and may have been well adapted for supporting the huge bulk of the animal. The feet of the sauropod are small (relatively speaking!), with short, stubby toes, yet animals that walk on soft mud tend to have spreading feet to distribute their weight more evenly. It is difficult to see how the compact feet of the sauropod could avoid becoming stuck fast in the soft, muddy bottom of a lake. If the dinosaur did, after all, live on dry land, then the long neck could have usefully functioned to allow the animal to browse the high foliage of trees (*see* p. 116). It has been suggested that an animal of the size of *Brachiosaurus* could not have breathed under water because of the pressure on the lungs. Their fossil remains seem to occur with other animals and plants, which are generally accepted as terrestrial. There is even some evidence from the tracks they have left behind that *Diplodocus* and its allies moved about in herds. They may have been the gigantic reptilian analogue of the elephant, and it may be no coincidence that the elephant also has its nasal openings on top of the skull, with the nostrils in this case sited at the end of the trunk – it has been suggested that some sauropods may have had a proboscis of some sort.

The balance of evidence seems to be swinging away from the original idea of swamp-dwelling giant dinosaurs to fully terrestrial habits. But other aspects of the dinosaur living habits are still more strongly debated. In the last few years a powerful controversy has arisen over whether the dinosaurs as a whole were cold-blooded, like all living reptiles (and there is no doubt that dinosaurs *were* reptiles), or warm-blooded, resembling mammals in this respect. Cold-blooded animals have to 'warm up' before they can be fully active; that is why lizards and snakes bask in the sun in temperate climates. For this reason they cannot cope with climates having greatly extended winters. Warm-blooded animals have the same body temperatures at all times, and can be more continuously active, but they use far more energy – and hence need more food –

One idea about the life habits of the dinosaur giants, published in *Joc and Colette at the Natural History Museum* in 1935

than cold-blooded animals of the same size. The posture of many dinosaurs, and particularly the carnivorous theropod dinosaurs, was fully erect with the legs beneath the body, and unlike the sprawling legs of living reptiles (*see* p. 116). The long back legs of such hunters look highly suitable for running, and as they did so the long tail may have been held erect as a kind of counter-balance (*see* p. 116). For any kind of prolonged activity, warm-bloodedness would have been a distinct advantage. Under the microscope even the bone structure of these dinosaurs looks more like that of living mammals than cold-blooded reptiles. On the other hand it can be argued that the sauropods like *Brachiosaurus* were *so* large, and with a relatively small surface area through which to cool compared with their enormous volume, that their cooling rate could have been slow enough to allow them to sustain more continuous activity than smaller, living reptiles. It has also been argued that the small mouths of the giants simply could not have downed enough food to support a warm-blooded metabolism, particularly plant food

that needs a lot of processing before it becomes available as energy. The armoured or plated dinosaurs include some species that seem too heavily burdened with protective armour to have been very active, but this does not necessarily mean that they had to be cold-blooded. Protagonists of the warm-blooded theory will show some of the horned dinosaurs charging over the Cretaceous plains like furious reptilian rhinoceroses!

The warm-blooded supporters tend to ally themselves with the supporters of a theory about the relationships of the dinosaurs. This is that the dinosaurs, or a close dinosaur relative, included the ancestors of the living birds. Some of the smaller, and most certainly highly active dinosaurs were about the size of a chicken, and there is more than a passing similarity between a running ostrich and the kind of reconstruction that shows fleet-footed, running dinosaurs. The point is that birds themselves are warm-blooded, like mammals, and if birds and dinosaurs are as closely related as now seems likely, then it obviously increases the likelihood that the dinosaurs themselves may have been warm-blooded. The pros and cons of this theory are the subject of much contentious debate by experts, but it would probably be correct to say that the greater number of specialists believe that at least the more bipedal of the dinosaurs, including small and large carnivores, were warm-blooded, active animals. A lot of the argument among the authorities on these animals is about the equivalence or otherwise of certain bones in dinosaurs and bird skeletons (and particularly the Jurassic bird *Archaeopteryx*); this makes for rather dry reading for the layman. But the end-product is the accurate reconstruction of the Jurassic and Cretaceous terrestrial world, over which the dinosaurs were indisputably dominant.

GRAPTOLITES – FLOATING COLONIES

In the 1830s and 40s geologists were beginning to unravel the mysteries of the early Palaeozoic rocks. The past world recorded in the rocks could not have seemed more alien. Many of the organisms that left fossil remains were of kinds now extinct, although some, like the trilobites, could obviously be placed into a phylum with many living representatives. The graptolites, however, were initially completely enigmatic; they were first described as plants! Their remains were so abundant that they could not be ignored. Their colonies, looking like so many miniature hacksaw blades, often completely covered bedding planes, and usually they were found in the absence of other kinds of fossils. It also became apparent that they could be useful in subdividing the intractable stretch of time from Late Cambrian to Silurian; they changed in obvious ways from one rock formation to the next. Ones with numerous branches seemed to dominate the earlier rocks, ones with fewer branches were later, while in rocks we would now recognize as Silurian and early Devonian, forms with but a single branch (or *stipe*) were abundant. With the discovery of better-preserved material it became apparent that the graptolites consisted of rows of tiny cups which were interconnected by a common canal – they were colonial animals. As noticed (p. 59) it is now known that the graptolites were an extinct branch of the phylum Hemichordata, an insignificant group today, consisting of a few encrusting colonial organisms.

How did these mysterious organisms live? In this case it is difficult to apply argument from analogy, because the graptolites

Left A modern reconstruction of the life habits of the giant sauropods, which could scarcely be more different from the 1935 version

Centre left The posture of a dinosaur compared with that of a living reptile

Bottom left The active, running posture of a dinosaur in the modern interpretation

are really rather different from any animals now alive. The modern hemichordates are all encrusters and virtually all colonial organisms (such as corals or bryozoa) are bottom dwellers in the sea. On this argument the graptolites should have been bottom dwelling-colonial organisms, living by filtering small particles of food from the water. Some of the graptolites (dendroid graptolites) were shrubby colonies, looking very much like living hydroids or some bryozoa. These were equipped with 'roots', and an inference of bottom-living habits seems eminently reasonable for them. But the great majority of the class, including all those with typical saw-blade profiles (graptoloids) which were widely used in the dating of rocks, certainly lacked any kind of rooting structure. When they were looked at in more detail it was found that the colonies grew from one single tube (sicula) which is often facing in a different direction from the tubes inhabited by the rest of the colony (*see* below). A slender rod (nema) usually grows from the top of the sicula. Since some of the graptoloids are large and robust, this slender rod seemed an inadequate basis for attachment to the sea floor, particularly in any sort of turbulent environment. Then there were some forms, like *Phyllograptus* (*see* p. 118), in which the colony grew over the slender nema to leave no visible means of attachment at all. It begins to look as if the graptoloids could have been free-floating animals, of a kind without any living counterpart.

Now the geological circumstances in which the most abundant graptolite faunas are found can be introduced into the argument. The most typical occurrence of these fossils is in dark, often sooty, black shales. In many localities graptolites are the only fossils to be found, but they occur in abundance. Sometimes they occur with cherty rocks that contain the remains of indubitably planktonic radiolaria. Some, at least, of the graptolitic black shales are what we can recognize, with the hindsight of plate tectonic theory, as the deposits laid down off the edge of the continental shelf, in a truly oceanic environment. These are frequently associated with volcanic rocks of oceanic type. So there seems little room for doubt that the graptolites *were* capable of living in an open-ocean environment. The fact that they occur in shales without other benthic remains makes it very likely that they were free floating, and this is consistent with the way some of the species were constructed. Dead graptoloids simply drifted down to the bottom where they were preserved by dark muds at a depth at which there was little oxygen in the water to support bottom-dwelling organisms. A few examples have been found where a number of graptoloids seem to have been associated together attached (by the nema) to a 'float'. Other species seem to have the nema extended into a kind of vane. There may have been some species which were attached to floating seaweed. Of course, many graptoloids also drifted into shallower water, where they are associated with a more normal kind of fossil assemblage. Like the pelagic trilobites, individual graptoloid species are very widespread, which is what one would expect of an animal with the wide ocean as its habitat. It is not surprising to find that the wide-

Below Growth and budding of a graptolite colony, enlarged about 20 times

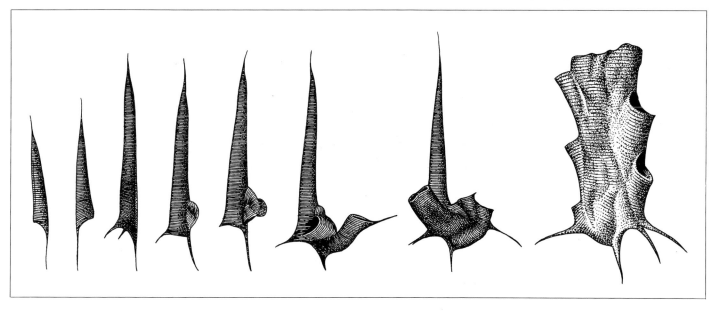

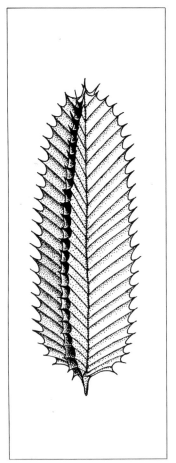

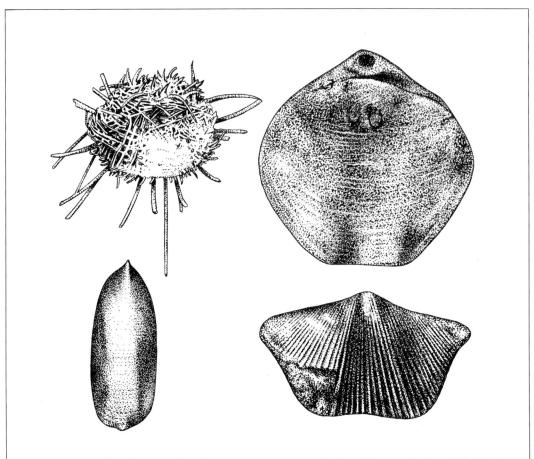

Above *Phyllograptus,* a graptolite colony with no means of attachment. About twice natural size.

Above right A sample of the different shapes achieved by the brachiopods: different shapes for differing modes of life

spread species have been used extensively to correlate rocks between different continents – they form the standard for the Silurian period, where the graptolite faunas are nearly the same all over the world.

In this case a combination of geological evidence and the evidence of the shape of the animal itself has been used to say something about how the animal lived. Many problems remain. Which way up did they float? Were there species adapted for living at particular depths? Why did graptoloids have an evolutionary trend from many branches to few branches? There are plenty of different answers available to these questions, but, like the metabolism of dinosaurs, no concensus has been reached as yet.

ANIMALS WITH TWO VALVES

Graptolites, trilobites, and dinosaurs are all extinct, and so in the examples we have considered so far there is no direct recourse to the study of living animals to help us bring our fossils back to life. Any analogies have to be with living creatures that are distantly related in the zoological sense, although there may be reasons to suspect that they have similar life habits to those that

the fossil animals once enjoyed. Some of the fossils with the simplest construction are those with the soft parts encased within two valves, principally the bivalves and the brachiopods. These two groups of animals are quite unrelated, but both still have living representatives in the oceans, and so it is possible to look at living animals to help with the interpretation of fossil examples.

Of the two, the bivalve molluscs are much the most significant in the sea today. The brachiopods are still numerous in some environments, but these tend to be rather inaccessible, such as in the deep sea, which inhibits their direct study in the field. Most brachiopods lead a rather uneventful life, attached by a stalk to a hard surface, like the underside of a rock, or another shell, filtering food from the surrounding sea water. There is nothing to suggest that the innumerable fossil brachiopods had any other method of feeding. Yet it is obvious that there are a great variety of shapes (*see* above) among the fossil forms, many of them unmatched in living species. There must have been many different ways for brachiopods to exploit their simple mode of life.

For a filter-feeding animal, it is sensible to

separate the inhalent current, bringing food, from the exhalent one, which carries away waste products. Brachiopod feeding is carried out by the *lophophore* (p. 61), a ciliated band usually carried on a loop, which also creates the currents used in feeding. Many brachiopods have developed a broadly wavy margin of the valves (*see* below), with the middle bent downwards. This form of the shell reflects the separation of the feeding currents from the exhalent currents. Cur-

Wavy valve margins in brachiopod. Note also the ribbing. Four times life size.

rents pass in through the sides of the shell, over the ciliated lophophore where the food is extracted, and then out through the depression in the margin of the valves. The lophophore is supported in several different ways. In some species the support is on a spiral made of calcium carbonate (*see* p. 120). This increases the length of the lophophore available to extract food from the feeding current, and so increases the efficiency of the process. Brachiopods begin to seem rather more complex animals than their simple shape suggests. Many species have a finely folded margin, crumpled like corrugated cardboard. Since the shell grows at the margins, a ribbed shell is produced, with fine or coarse ribs according to the species. A ribbed shell of this kind is often stronger than one without ribs (for the same reason that corrugated iron is used for roofing), and it is probably no coincidence that ribs may be developed on bivalves too. But the wavy margin may serve another function. For one thing it increases the *length* of the margin, allowing the animal to admit more food per unit of length. The wavy margin also prevents the entry of irritating particles of sand, with which the animal cannot cope. However, there is still a problem at the crests of the corrugations, where the gap between the valves is larger; many brachiopods solve this particular problem by having fine, hairlike spines which cover the crests of the corrugations, and so prevent the intrusion of unwanted particles.

Although most brachiopods were attached to a firm substrate by a stalk which emerges through a hole in the pointed end of the animal, there are some in the fossil record in which there is no opening for such a stalk. These must have lain freely on the surface of the sediment. In some of these species one valve is greatly thickened. This is likely to have been the lower valve, and fossils are sometimes found in 'life position', which confirms this. The other valve sits in the thick one like a lid, and the whole animal is curved so that the margins of the valves were kept clear of the sediment. The end result is an animal that looks very much like some kinds of oyster, although a glance at the internal feeding structures shows at once that they are brachiopods, unrelated to the bivalves they superficially resemble. Some species had long spines (anchors?) attached to the lower, thicker valve.

Unlike the brachiopods, many of the modes of life of living bivalve molluscs can be more or less directly matched in fossil

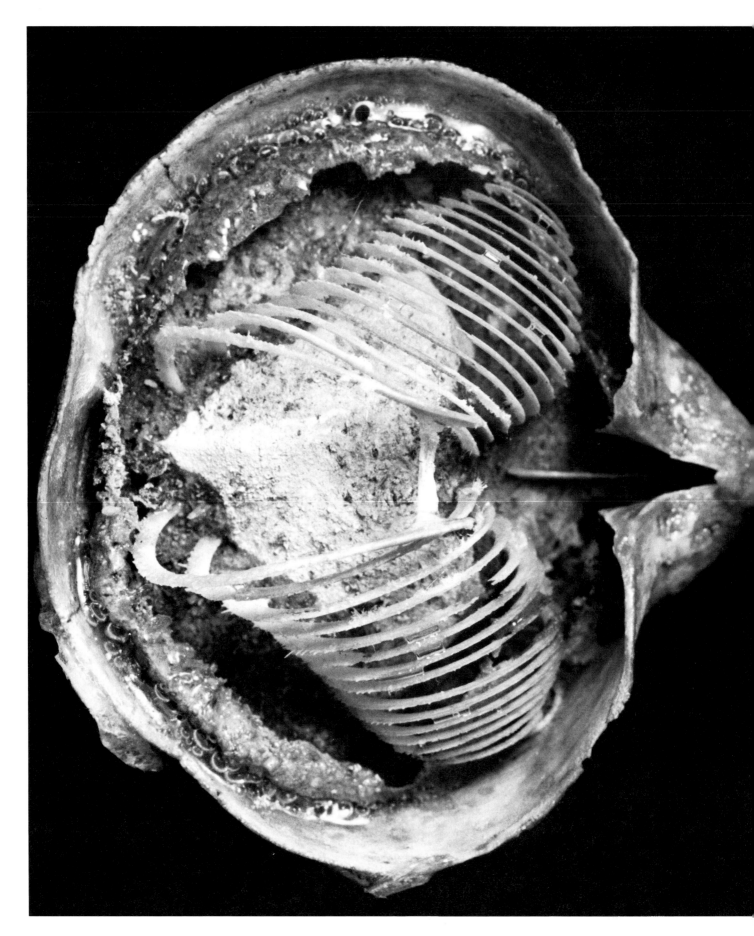

Above Swimming in bivalves: living *Chlamys* escaping starfish predators

Right Burrowing bivalve mollusc in life position; the siphons retain contact with the surface

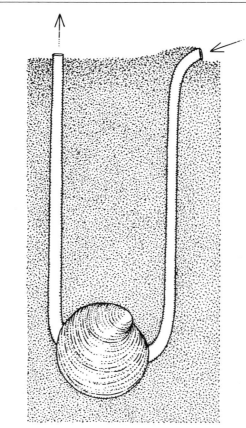

Left Spiral supports for the brachiopod feeding system, preserved inside one of the valves, and etched out with acid

examples. So by studying how the life habits of living bivalves are reflected in the shapes of the shells we can make deductions about how the fossils lived. A major difference between brachiopods and bivalves is that the latter are capable of moving freely, using their foot to crawl or dig. Of course, some bivalves, like mussels and oysters, *do* remain fixed throughout their adult life, and it is possible that these fill the ecological role to-day of some of the large brachiopods of the Palaeozoic. Some of these lived in shallow water sites and were gregarious like mussels. But since brachiopods were unable to do many of the things that bivalves do very efficiently, it would be unwise to attribute the decline of the brachiopods to diversification of the bivalves.

Burrowing bivalves, which are typically found in the soft bottoms of shallow seas today, dig themselves down into the sediment, where they escape the attention of many predators (except birds with long bills!). Many of them retain contact with the surface of the sediment by means of long *siphons* (*see* left), which enable them to breathe and feed on small organic particles. Most bivalves with siphons develop a gape at one end of the valves, so that they do not close entirely in the region from which the siphons protrude. It is easy to recognize such gaping valves in fossils, and there are accompanying changes on the internal structure of the shell which are associated with species with long siphons. Hence bivalves as old as Palaeozoic can be identified with some confidence as ancient burrowers. It can be confirmed by discoveries of fossils preserved in their life position. Often the process of 'digging in' is assisted by a characteristic pattern of chevron-shaped ribs on the surface of the shell, which are also obvious features of some fossil shells. Such V-shaped ribs are not found on brachiopods, which have never developed the capacity for burrowing.

Other kinds of bivalves have used their capacity for free movement to the utmost by becoming efficient swimmers, clapping their valves together like castanets to move through the sea to find new feeding grounds or to escape the threat of predators. These bivalves have fan-like shells, often strongly ribbed (*Chlamys*). While the bottom-dwelling or burrowing bivalves mostly have a pair of strong muscles to pull the valves together, in the swimming forms the muscles have become modified so that there is one particularly powerful muscle, centrally placed, to produce the powerful clapping movement that propels the animal through the water. This muscle leaves a conspicuous scar on the inside of the shell, an impression which is easily preserved in the fossil state. So it is possible to deduce swimming habits in fossil bivalves by comparison with the living forms. These swimming habits go back into the Palaeozoic, to the Carboniferous or earlier. There are a few bivalves which are found in rocks deposited in the same sort of black shale environment that was mentioned above in the discussion of graptolite life habits. Some experts maintain that these molluscs were able to swim in the open ocean, or that they were attached to floating seaweed: some such mechanism has to be invoked to explain how these bivalves came into an environment that lacked bottom-living animals.

I have only selected a few examples of deducing modes of life from these fossils with two valves; the examples could be multiplied many times. The most important point is that relatively complicated life habits can be inferred from a careful consideration of the anatomy of even the simplest looking animal, particularly when there are related, similarly adapted animals for comparison.

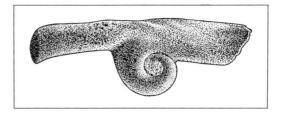

A palaeontological enigma: *Janospira*, a small Ordovician fossil of unknown relationships

PALAEONTOLOGICAL ENIGMAS

Every now and then the fossil record throws up fossils which are palaeontological puzzles. They are obviously the remains of some kind of animal, but the problem is to decide what kind. They tend to be rather rare, preserved in a special way. And, of course, like so many palaeontological matters, they stir up arguments between specialists who think they have a way of solving the enigmas.

Some of these puzzling fossils are quite small. A few years ago a minute fossil only about 2 mm long was recovered from limestones of Ordovician age, and christened *Janospira*. It looks remarkably like a trumpet. The 'mouthpiece' and the 'horn' of the trumpet are both open, and a coil hangs down from the middle. The end of the coil is closed, and it seems reasonable to assume that the animal started growing as a coiled shell. Then it must have changed its mind, and started to grow in two directions, a narrower tube into the 'mouthpiece' and a broader one into the 'horn', and there is some evidence that the 'horn' end continued to get wider and longer. The problem is that it is hard to equate this kind of growth with that of any known animal group. It *looks* like some kind of mollusc, but no mollusc fits easily into this pattern of growth; various people have suggested it might be some kind of snail, or perhaps a monoplacophoran (*see* p. 76). It remains a puzzle. It is found along with fossils of planktonic organisms and it is possible to explain the change in growth between coil and trumpet as a change that happened when the larval shell settled on the bottom. But since virtually all phyla of animals have planktonic larvae, this is no help in solving the puzzle.

Some larger fossils are even more puzzling. A number from the famous Middle Cambrian Burgess Shale of western Canada are so strange that they have even been claimed as the representatives of 'extinct phyla'. Certainly they are difficult to accommodate comfortably in groups alive today. One of these, *Hallucigenia* is shown opposite. This curious animal has what was probably a gut and a set of tubules arising from it; below this are curious pointed structures. What can it be? One suggestion is that early on in metazoan evolution there were a number of experimental designs which were not to give rise to direct descendants; naturally these do not fit into the pigeon holes based on living animals. *Hallucigenia* may have been one of these. There are problems with designating these as new phyla, not least that nobody can define a phylum objectively, and my own view is that, marvellous though these animals are, we should still try to relate them to other organisms (fossil and living). *Hallucigenia* would defy all such attempts at the moment.

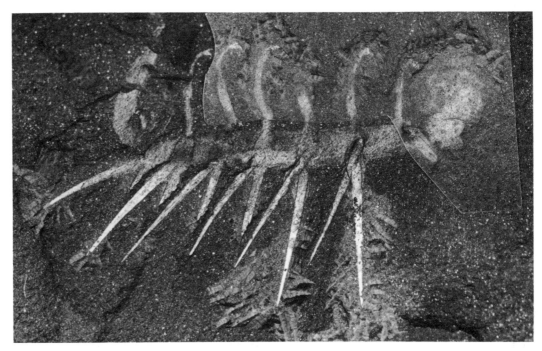

CHAPTER SIX

EVOLUTION AND EXTINCTION

'Dead as the dodo'. The finality of the phrase sums up a general attitude towards extinct animals. They had their day, were somehow unfit for the modern world, and as a result, died out. This chapter explores some of the causes of extinction, to show that without extinction there would have been little chance for evolution. While the attitude of conservationists to prevent the wanton destruction of species is wholly laudable, the fact remains that in the normal course of events species *will* become extinct. The fauna of the present day is the product of repeated changes in the faunas of the past, and there is no reason to suppose that the process will stop now that *Homo sapiens* covers much of the world, although there is equally no reason why he should speed up the process by wholesale destruction of habitats or excessive hunting. As the world has changed, so have the faunas and floras, with evolution and extinction playing their complementary parts in constantly re-shaping the biosphere.

There are really two types of extinction. In one, the process is as final as in the case of the dodo. A species dies out completely, leaving no progeny, no descendants. If a whole group of related species become extinct at about the same time a major animal group may become extinct. There are no dinosaurs lurking in remote corners of the world, not even in Loch Ness. More than anything else it is the disappearance of such major groups that has changed the appearance of the fauna through geological time. A second type of extinction is involved in the generation of new species. One species gives rise to another, by any one of a number of processes, and ultimately the parent species may become

extinct. But in a sense its genetic material lives on in the daughter species. The only totally 'dead' groups are the side branches of the evolutionary tree (*see* below), for which the long chain that connects the living species with the ancestors from which they sprang has been irrevocably severed.

If extinct species, which are the ancestors in distant geological periods of the living fauna, have hard parts and are readily fossilized, then, it should in theory be possible to

Extinction by transformation, as opposed to extinction by termination of a fossil lineage. The mammals on the left include many extinct fossil species, but also have living representatives; the trilobites on the right are completely extinct

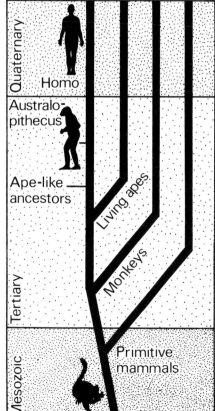

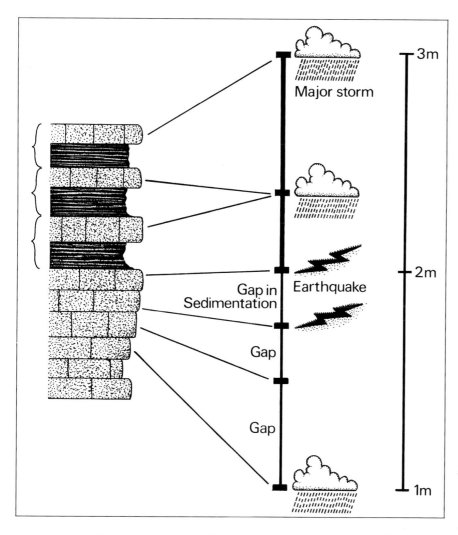

Breaks in the fossil record. The rocks accumulating over this three million year period may only do so during exceptional events like storms or earthquakes; not *all* time is recorded in the rocks

read the past simply by collecting backward through the strata from the youngest rocks to the oldest. Anybody who tried to collect the evolutionary story directly from the rocks would be disappointed. As Darwin was well aware, the rocks seem to betray many gaps and holes in the record. Somehow the ancestor hardly ever seems to be sitting there, where it should, in the rocks immediately below the descendant species. Sometimes a species which looks in most of its features as if it should be ancestral to a whole group of animals turns up in surprisingly young strata. The fossil record is imperfect and capricious. This has led some palaeontologists to reject the stratigraphic order in which fossil species occur as being of no significance at *all* in the elucidation of the evolutionary process. It is certainly not surprising to find primitive-looking animals in rocks younger than we might expect; after all, the 'living fossils' which are so important in unravelling the relationships between different animal and plant groups are all in a sense organisms that have out-

lived their time. Some primitive animals are well-adapted to their own ecological niches from which they have never been displaced. Evolutionary events do not happen in a regular, step-wise fashion; it is a rather messy, irregular process, with survivors persisting alongside novelties, momentous evolutionary steps being taken by humble organisms, while the dominant animals may undergo spectacular evolutionary extravagances that are doomed to total extinction. Some theorists have even said that almost the whole pattern of evolution may be accounted for by chance alone.

Why does the fossil record seem to be so imperfect? Part of the reason, and the one that used to be cited as most important, is that there are many breaks in even the most continuous-looking rock succession. Bedding planes are the record of such breaks. In some successions deposition only occurs irregularly, as after a major storm, and estimates put the rock as representing as little as 1% of the total time in some sedimentary sites. Even so, the evolutionary process is relatively slow, and one might expect to find the evolutionary story visible even through the small pieces of the record just as you can recognize the subject of a *pointilliste* painting by standing at a distance. A more important reason is that most rock successions record shifts in *facies* (p. 38): the sedimentary (and biological) environment changes with time, and as this happens the evolving animals are carried elsewhere and the descendant species are recovered from a different rock succession, maybe many kilometres away.

But perhaps the most significant factor stems from the way new species are derived. Where the generation of new species has been studied among living animals a most important cause of the splitting of a new form is the isolation of a population of a species at the fringe of its range. These fringe populations become separated, change in response to some slightly different set of conditions and eventually become different enough from the parent stock so that, even if given the chance, they will not interbreed – in fact, a truly independent biological species. A shift in conditions may allow the descendant species to spread out and even displace its progenitor. Because the inception of a new species happens at the edge of a population, the chances of finding the actual site where it happens preserved in the rocks is rather low: the point of origin is nearly always 'somewhere else'. The result in any particular rock section is something like what we actually see. This is a rather patchy mosaic of species that all seem to be related, and in some cases may record actual ancestors and descendants, but which can certainly not be read 'like the pages of a book'. Added to this is *another* kind of change, which can be recorded in sedimentary successions. This is a slow drift of change which occurs within a species through time, a shift in shape perhaps, or an increase in size, which cannot be explained by isolation of populations. In this case – and its causes are still rather mysterious – the ancestral species is transformed into the descendant one and so does not really become extinct in the normal sense of the word. Some palaeontologists hotly dispute that this kind of change represents evolution at all, and there is a lot of hair-splitting about whether or not the descendant is 'really' a different species. But regardless of the theoretical arguments, there are many rock sections in different parts of the world where this kind of change has been seen. It is of practical importance, too, because these kinds of small changes often occur in the commonest fossils, and become the basis of stratigraphic zones for the precise dating of rocks by their fossil content.

This complicated introduction is necessary to understand the complexity of the fossil record. For example, if the new species arose by the kind of isolation of marginal populations that is important today, then we might almost *expect* the fossil record in any limited area to have gaps and jumps in it. In fact, the fossil record might not be so incomplete as has been thought; the jumps may be a natural phenomenon. The generation of new species, and the extinction of others has been a continuous process since the Cambrian. If one considers life in the seas, there was a steady growth in the total number of species during the Cambrian (*see* above right), and the number has remained high since then. But there have been two important times when extinction greatly exceeded the generation of new forms, and these were the times when whole groups – like the rugose corals or the ammonites or the dinosaurs on land – passed from the world for ever. Very important extinctions occurred at the end of the Permian and the end of the Cretaceous periods, and it is no coincidence that they also define the end of the Palaeozoic and Mesozoic, respectively.

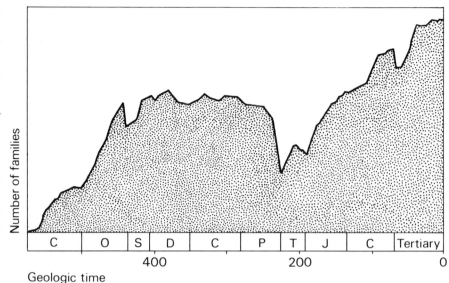

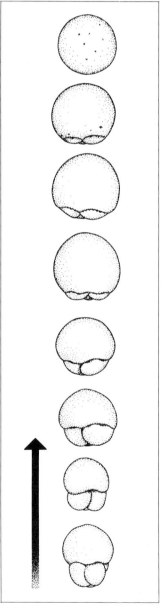

Above Numbers of major groups of marine organisms through geological time. After an early climb through the Cambrian, the numbers are rather uniform except at the periods of major extinction, especially at the Permian–Triassic boundary

Left Gradual transformation of the foraminiferan *Globigerinoides* into *Orbulina* (Miocene). Photograph of *Orbulina* (*below*) × 200.

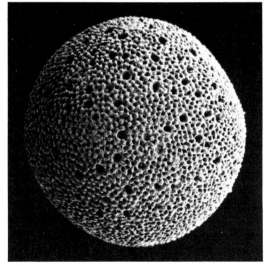

They were the great biological crises that changed the faunas of the earth. Here we have to invoke more powerful causes than the normal processes of change described earlier in this chapter, and I shall return to them later. But first it is necessary to describe the kind of small scale evolutionary changes that are the bread-and-butter of the fossil record.

THE MOSAIC OF EVOLUTION

The evolutionary story is best preserved in the least conspicuous of fossils. Unfortunately, the record has not revealed the step-by-step changes that led (for example) to the evolution of birds from reptiles. Such momentous changes have taken place in sites where the fossils stand a low chance of being preserved. To see the record at its fullest one has to turn to sites where there is every reason to suppose that sedimentation has been continuous. One such place is in the deep sea, where the steady rain of plankton carries on regardless of the shifts in sea level that affect the continents, and the sea bottom is a graveyard for the rain of tiny organisms that die nearer the surface. Some

of the small planktonic foraminiferans (p. 126) seem to record the kind of continuous changes that evade the palaeontologist elsewhere. Minute species of the genus *Globigerinoides* change step by step into a perfectly spherical form known as *Orbulina* (*see* p. 126). The appearance of this spherical animal serves to define the base of the middle Miocene period. The same sequence of changes from the one species to the other has been found repeatedly in many sites all over the world, and one has to visualize the slow transformation of the chambered shell into the spherical one. The cause of the change is a mystery: but whatever the reason the effect is of the greatest use in dating rocks of Tertiary age.

This kind of simple change can be matched by other examples from the planktonic foraminifera. It seems to be characteristic of planktonic organisms in general. Graptolites also show continuous changes of this kind. In the earlier half of the Ordovician, slender 'tuning fork' graptolites of the genus *Didymograptus* (Colour plate 10) show a progressive change to stouter, longer forms, while the v-shaped species of *Isograptus* produce longer, more robust kinds of colonies (*see* left). Ammonites also tend towards a continuous sequence of changes, sometimes with the multiplication of ribbing or with progressive changes in the outline of the shell (*see* p. 128). In these planktonic or pelagic organisms it seems that whatever advantages were conferred on the species by the change were transmitted to the population as a whole.

Among bottom-living fossils it is perhaps more usual to find rather sudden jumps between one species and the next, probably because the species evolved by geographic isolation in the manner described previously. The sea urchin *Micraster* is one of the commonest fossils in the European chalk (Cretaceous), where its heart-shaped tests have acquired the common name of 'shepherd's crowns'. The chalk seems to have accumulated as a pure, lime ooze, largely formed by the microscopic remains of minute algae, and foraminifera; although not a sediment of the open ocean, its record of deposition is remarkably continuous. Changes in the heart urchins record an evolutionary story connected with burrowing habits. As the sea urchins acquired the habit of burrowing deeper in the soft, chalky sediment their tests underwent a whole series of small

Gradual changes in graptolites through time: *Isograptus* (Ordovician)

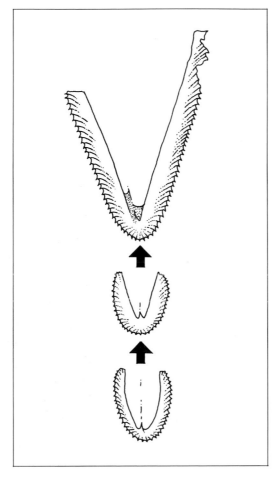

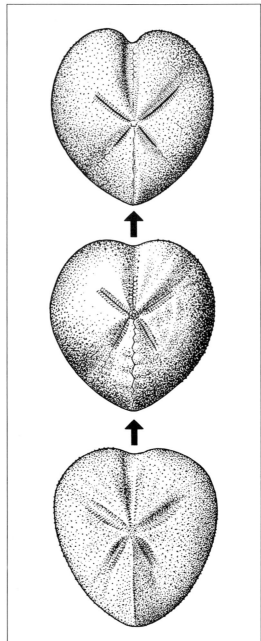

Left Changes which occurred in the heart urchin, *Micraster*, during the Cretaceous

Far left Changes in ammonites through time: *Cardioceras* (Jurassic)

changes: the entire animal became higher and deeper, the little plates that make up the petal-like ambulacral area (p. 81) became more inflated, and the little lip around the mouth became exaggerated (*see* above). The series of species leading from *Micraster corbovis* through *M. cortestudinarium* to *M. coranguinum* record these changes, and the same species may be collected in the same order across much of Europe. They have become fossils of importance in determining the zones of the chalk. In this case it is possible to correlate the changes in the fossils to their mode of life by comparison with living sea urchins, which show similar changes in relation to life habits. As the animals learned to burrow more deeply it became necessary to retain contact with the surface for breathing through a long tube, kept clear of sediment by extended tube feet. The points of attachment of these specialized tube feet can be observed to become modified in a special way as the tube feet become longer. The *Micraster* species were not the only organisms to show slow, but ordered, evolutionary change as the chalk was slowly deposited; the large molluscs of the genus *Inoceramus*, as well as the tiny foraminiferans that go to make up much of the sediment also changed, and all of their changes act as a chronometer for the passage of Cretaceous time.

128

The chalk is an exceptional deposit in its continuity and uniformity, without the kind of drastic change in sedimentary facies that introduce the snags into more usual sedimentary sequences. Even the chalk is not without its changes – there are beds where deposition temporarily slowed down or ceased, called 'hard grounds' because the sediment surface became crusty. Here a whole series of animals that are normally rare in the chalk become numerous, and unlike the sea urchins their evolutionary history cannot be deduced from careful collecting in the beds above and below, simply because they cannot easily be found there. With many fossil groups this is the normal state of affairs; fossil remains are found at certain special horizons, some of which become justly famous for the beauty of their preservation. They tell us much about the animal species concerned, but nothing about the history of the animal in the time immediately before or after. Usually the gaps between the fossil-rich horizons may be measured in millions, if not tens of millions of years, and they are often widely separated geographically. It would be a mistake to think we could stack these fossils up in their time relationships to give us a picture of how they changed, in the manner of the sea urchins. There are too many ambiguities in the 'gaps'. The best that can be done in these circumstances is to look very carefully at the features they show, decide which ones were derived in the course of evolution and which ones are primitive, and construct a diagram to show the progressive stages in their pedigree (technically, a cladogram). As usual, it is a matter of debate how imperfect the record of a particular group of organisms is, and different scientists have differing views about how important the stratigraphic order should be.

HUMANS

'The proper study of mankind is Man.' The resources put into the hunt for the fossil remains of humans are proof of Alexander Pope's aphorism. The search for palaeontological evidence for human derivation has occupied more newspaper space – and provided more scandal – than the rest of the fossil firmament. Since Darwin published *The Descent of Man* in 1871 the idea that humans and the higher apes were closely allied had lodged in the common consciousness. Definitive fossil proof was lacking until the recent decades. The idea was prevalent that there would be a 'missing link' lodged in the rocks – a single perfect amalgamation of human and ape characteristics that would prove our ancestry at a stroke. Such a fossil resolutely refused to turn up. So eager were the scientific community for such a discovery that when the Piltdown Man skull was 'unearthed' in 1912, it was quickly accepted by many respected workers as the answer to the missing link puzzle. Here was a skull that really seemed to display an amalgam of ape and human characteristics. Sadly, that is exactly what it was, an elaborate forgery composed of bits of human and bits of ape, carefully matched to look genuine. It was exposed as a fake when the different ages of the different pieces were proved, and, in particular the recent ages of the human and ape 'fossils' were demonstrated by radiocarbon dating. Even now controversy still rages over who was 'in the know' about the forgery.

Gradually, though, other genuine pieces of human-like apes (or ape-like humans) began to turn up from areas well-removed from Europe, notably China, Java and Africa. It has been the African continent more than any other which has yielded the fossils that have provided not one missing link but a whole chain of intermediate forms, showing that the evolution of humans was much more complex than had been supposed. New discoveries of fossils of *hominids* (Hominidae is the family that includes humans and their ancestors) are being made every year, now that the kind of sites occupied by our distant ancestors have been identified in the deposits formed in and around ancient lakes, rivers and in caves. Many of the exciting discoveries of the 50s and 60s were made by Louis S. B. and Mary Leakey, and the family tradition has been continued by their son Richard. Opinions about the meaning of the fossils seem to have changed with every subsequent discovery, but gradually some sort of consensus is emerging.

Important finds have now been made in many localities, notably in Olduvai Gorge and Laetoli in Tanzania, and in South Africa, Kenya, and Ethiopia. Some of the ape-like hominids which were initially supposed to have been our very ancestors have now been recognized as a side branch in our history: most of these are now placed in the genus *Paranthropus*. The skulls of the robust species of this genus are flat fronted, and the brain case is obviously much smaller than in any human, but they lack the prominent canine teeth that make the gorilla such a fearsome-looking animal, and in this respect are more like our own ancestors. They have been found in deposits as old as 2.5 million years (dated from radioactive minerals) and up to about 1 million years old. The other generally accepted early hominid genus, *Australopithecus*, includes the species *Australopithecus afarensis* (a remarkably complete skeleton of which was christened 'Lucy') which is believed to represent the common ancestor of humans and chimpanzees.

Younger *Paranthropus* and *Australopithecus* species may have been contemporaries of our early ancestors. *Paranthropus robustus* may have been adapted to a dry, savannah type of climate. A more slender species of *Australopithecus*, *A. africanus*, which is found in deposits even older than

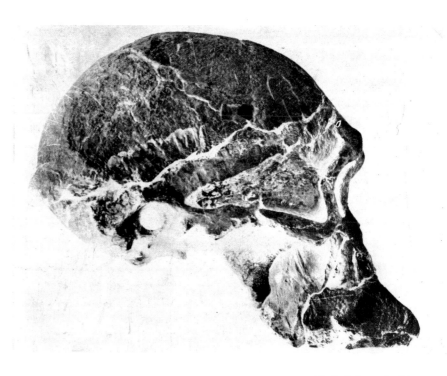

Above Skull of *Australopithecus*; a close relative certainly, but probably not the direct ancestor of modern humans

Homo habilis (left) and *Homo erectus* (far left)

130

those with the robust species but overlaps with these in time, is often considered to be closer to the main line of human evolution, but its place as the ultimate ancestor of our living species is still open to debate. Whatever the position of *A. africanus*, it is clear that by about 2.0 million years ago the first species that can be placed in the human genus, *Homo*, had not only evolved, but was manufacturing tools for complex tasks, and probably living together in small communities, and so was displaying the kind of co-operative habits we like to associate with ourselves. This species is known as *Homo habilis*. Although *H. habilis* does not have the kind of cranial capacity typical of modern humans, its brain was larger than that of the australopithecines (this does not necessarily mean that it was very much more intelligent). Some remains of limbs that have been found are much more like those of humans than those of either apes or *Australopithecus*, but there has inevitably been the usual debate about whether it is *the* ancestor, or a first cousin on a side branch, and even whether the remains all represent a single species. But certainly the *H. habilis* material is more closely related to us than are the australopithecines. Up to this stage the history of the family of humans has been recovered from fossils confined largely to the African continent.

Homo erectus, by contrast, seems to have been much more widespread. Remains of this species have been found in China, Java and Europe as well as in Olduvai in Africa. In age they more or less follow those of *H. habilis*, falling in the range from about 1.7 to less than half a million years old. The most likely explanation of the widespread distribution of *Homo erectus* is that the species spread from the African continent over a large part of the Old World, moving up into temperate latitudes. This must have provoked new responses from the dispersing populations. It is certain that some *H. erectus* populations regularly used fires, and others developed more sophisticated stone tools, including hand axes. Known fossils of *H. erectus* from Africa are older than those from the rest of the world, which is what one might expect if the invasion idea were correct, but one always has to remember how rare the fossil remains of humans are, and it only requires one fossil in the wrong place at the wrong time to upset the most cherished theory. In shape, too, *Homo erectus* is midway between *H. habilis* and modern humans for most (but

not all) of its features. The brain case is lower than ours, and there is a distinct ridge over the brow, and one at the back of the head, too. Some palaeontologists put more stress on the differences from modern humans, and there is, as the reader would by now anticipate, the debate about whether *H. erectus* is a side branch in the story. But it is clear that if one statistically treats all the features of *H. erectus* and what are often called the early remains of our own species, *H. sapiens*, then there is an almost perfect gradation between the one and the other.

At about the same time as the small, but crucial changes between the species of *Homo* were taking place, the climate of the northern hemisphere was passing into the coldest phases of the Pleistocene Ice Age. In 1856 a skeleton was discovered in the Feldhofer cave in the Neander Valley, which has since given its name to a suite of early human fossils known as 'Neanderthals'. Most of these fossils are about 50 000 years old. The typical Neanderthal has a skull that bulges out at the sides, and with a lower cranial shape than modern humans, but the capacity of the brain case matches (or even exceeds) that of living populations. The bones of the rest of the body are generally stouter than ours. Skeletal remains of this kind are widely distributed across Europe and extend into the Middle East. These individuals have been interpreted as specialized to cope with the cold conditions present when the remains of Neanderthals are commonest – at the beginning of the last glaciation. Even today the limbs of Inuit (Eskimos) are relatively shorter than those of Masai tribesmen from the tropics, and the Neanderthals were similarly proportioned to Eskimos, although considerably more muscular. There are fossil populations with ages around 300 000 years that are broadly transitional between *Homo erectus* and Neanderthals. So the Neanderthals may have been a specific race adapted to the severe cold climate of the glacial period, and notably confined to the colder habitable parts of the western world. Near the end of the last glacial period they may have been displaced by more modern, or in places became people modified (e.g. by interbreeding) to become more similar to modern humans. Skulls indistinguishable from those of typical living *Homo sapiens sapiens* (Cro-Magnon) are present at about 30 000 years before the present, with extensive but much less complete evidence extending back to 100 000 years in Africa and the Middle East. Some of these people had all the behavioural characteristics of modern humans, including the propensity for burying their dead – which is why fossil remains become so much more common in the Middle and Upper Palaeolithic. The use of tools goes back to *Homo habilis* and probably beyond – even chimpanzees are quite capable of thinking of the possibilities of pointed sticks – but their use and design no doubt became progressively more sophisticated: hence the recognition of the old divisions (time-cultural divisions based on a stratigraphy of stone tools rather than fossils) of Palaeolithic, Mesolithic and Neolithic. It becomes a little difficult to *define* modern man. Our complicated patterns of speech are characteristic but the vocal modifications necessary for speech are not readily recognized from bones. The first complex ceremonial centres and cities ('civilization') were established by about 9000 years ago, which is, geologically speaking, in the twinkling of an eye from modern humans' appearance over 50 000 years ago. All the human races today are usually included in the single subspecies *Homo sapiens sapiens*. Our subspecies completed the dispersal of the genus *Homo* which began with *H. erectus*. Fossil evidence shows that humans moved into the New World, via the Bering Straits from Asia by about 15 000 years ago, and possibly into Australia 40 000 years ago. It is probably this wide geographic dispersal, and the very different conditions this particular species meets over its range, which is responsible for the wide variation between races. The adaptability on a cultural scale is unique

Reconstruction of Neanderthal man

Three different hypotheses relating humans and our close relatives which anthropologists are debating at the moment

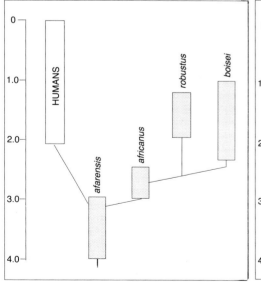

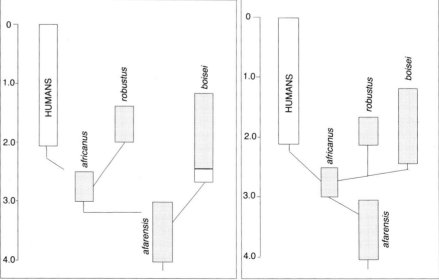

to humans, and perhaps his best species diagnosis, but it is a product of his mind, and not of his fossilizable parts.

Sadly, *Australopithecus* species did not survive the Pleistocene. It would be exciting to think of them lurking in some remote corner of the world – perhaps as the progenitors of the legend of the Yeti – but it is difficult to believe that any large land animal could have escaped the scrutiny of the twentieth century, unless he were equipped with the intelligence of *Homo sapiens*. Whether our own ancestors were involved in the demise of our nearest relatives is a matter for speculation.

TIMES OF MAJOR EXTINCTION

Whether hominid or sea urchin, the kind of evolution and extinction we have described so far has been on a relatively small scale. Splits (or *dichotomies*) have separated different lineages (for example, *Homo* and *Paranthropus*) one of which may become extinct, but there is no question of the whole of the primates, the group which includes the monkeys, apes and lemurs as well as man himself, being threatened with extermination. Even the traumas of an Ice Age, which produces races or species to cope with cold conditions (of which woolly mammoths and neanderthals were probable examples), do not necessarily stimulate the mass extinction of whole groups of animals. Yet the extinctions that occurred at the end of the Permian and again at the end of the Cretaceous periods were of this kind. These were times of crisis for the whole fauna (and flora) through which only a few groups passed unscathed; many major groups of animals passed from the earth for ever at these crises (*see* p. 135). In some cases the extinction was preceded by a general decline in number of families, so that the trilobites had already been reduced to a few forms by the Permian, and their removal before the Mesozoic was only the *coup de grace*. But in other cases the termination was more abrupt – the classic example of this is the extinction of the dinosaurs at the end of the Cretaceous. These events seem enormously destructive and yet they were also to allow the subsequent proliferation of other groups: extinction and evolution were partners in shaping the modern biological world. While the dinosaurs occupied the dominant places in the terrestrial ecosystem there was no obvious opportunity for mammals to become the prevalent large herbivores, but once the dinosaurs were removed the rich grazing of the Tertiary was there for exploitation. In both the Triassic and the early Tertiary there was a short time lag after the preceding extinction events to allow recolonization of the available ecological niches, but it was astonishing how quickly the faunas recovered to something like their previous diversity. The two major extinction events are the most pronounced of a larger number of phases when extinctions were apparently higher than normal. At the end of the Triassic, for example, almost all the ammonites that had flourished in that period became extinct, a mere two groups surviving to give rise to the hosts of different kinds that populated the Jurassic seas. At the other end of the geological column, the end of the Ordovician saw the extinction of many of the characteristic trilobite families that make Ordovician fossil assemblages easy to recognize, and the same event affected brachiopods, nautiloids and graptolites. These high rates of extinction seem to define the ends of many of the geological periods, indeed, it was probably the different gross characters of the faunas on either side of the boundaries that enabled the early geologists to recognize the different major periods to start with. The events at the end of the Palaeozoic and Mesozoic were still the most dramatic. It has been suggested that the extinction of animal groups at these times may have been because of lower rates of *evolution* rather than particularly high rates of extinction, such that the normal rate of replacement of extinct animals by new species just did not take place: in any event the effect is the same.

At the end of the Palaeozoic the whole group of calcite rugose corals became extinct, to be replaced by the aragonitic Scleractinia in the Mesozoic. Similar drastic changes occurred in the Bryozoa. Most of the typical Palaeozoic brachiopods disappeared, or were drastically reduced in numbers, and only two major groups, the rhynchonellids and terebratulids, prospered in the Mesozoic. Among the crinoids, which were extremely varied in the Permian, only one genus survived into the Triassic. The blastoids disappeared before the end of the Permian. Nautiloids declined, and the end of the period was a testing time for ammonoids and other molluscan groups. Trilobites and other Palaeozoic marine arthropods became extinct, and the Mesozoic really marks the inception of most of the important arthropod groups, like the crabs and lobsters that dominate the oceans today. Among terrestrial animals there were also profound effects. Most of the groups of reptiles that dominated the Permian did not survive into the Triassic, though some of these were more or less ancestral to later forms. The great radiation of the archosaurs, including the dinosaurs and crocodiles, dates back to the Triassic and just into the top of the Permian. Plesiosaurs and ichthyosaurs have a similar origin in the Mesozoic. Even many important groups of plants, like the cycads, have a history that roots back into the Triassic, but not far beyond, and the end of the Palaeozoic marked the decline, if not the end, of the great lycopod trees that had been such a feature of the Carboniferous coal swamps. In short, almost every group of animal or plant was affected by the late Permian 'event', and the resulting regeneration after the extinction laid the foundations for the modern fauna.

At the end of the Mesozoic, in the last part of the Cretaceous period, the dinosaurs died out, apparently suddenly. Their dominance over terrestrial habitats that had lasted for more than a hundred million years came to an abrupt end. In a habitat that could scarcely be more different, the open sea, the ammonites also died out, leaving no descendants. The flying lizards – pterosaurs – also took to the air for the last time in the late Cretaceous, and the marine plesiosaurs came to the same end as the ammonites. Even the tiny planktonic foraminifera underwent a revolution at apparently the same time. An event of such magnitude, affecting a diverse selection of organisms in many different habitats, demands an explanation. And there is no shortage of different theories: sometimes there seem to be almost as many explanations as the number of major groups that became extinct! In the welter of such hypotheses it is important to hang on to the facts, and not get carried away by the most compre-hensive and dramatic explanations. It is equally important to remember that certain groups of animals certainly did *not* become extinct at the same boundary, so at the end of the Cretaceous the lizards and snakes (both small) and the crocodiles (distinctly large) 'came through' apparently unscathed, as did the mammals, which were to inherit the roles vacated by the dinosaurs. So we cannot simply bombard the earth with some sort of fatal dose of radiation, unless we grant some rather exceptional properties to some unexceptional organisms. Also, since the last dinosaurs and last ammonites occur in different kinds of rocks, the one terrestrial, the other marine, there is always the problem of correlating the different events; it is hard to be certain that the extinctions occurred at *exactly* the same time.

Not surprisingly, the death of the dinosaurs has generated numerous ex-planations, but it is a pity that most of these theories have paid little attention to the drastic changes that were going on among all the other groups of animals and plants. One theory has it that the climate may have become too dry to support enough vegetation to keep the giant herbivores alive; after all, if the herbivores *had* died out, the carnivores that preyed on them would have become extinct automatically. But there is no real evidence of the sort of widespread desert conditions necessary to induce such a change, which should be reflected in the rock deposited at the Cretaceous–Tertiary boundary, and this explanation could not account for the extinction of the other groups (particularly in the sea!) or the survival of the crocodiles. If not too dry, then perhaps the climate became too cold, so that the dinosaurs starved to death during the long inactivity of the interminable winter seasons. This period was not one of glacial activity, sadly for this theory, although there was a climatic cooling at the time, and anyway the effects would have been cushioned in the sea, where events almost as dramatic were taking place. What about change of diet? The Tertiary rocks are dominated by flowering plants, and so is it possible that the dinosaurs could not cope with this new kind of sustenance? In this case the timing is quite wrong, for the big evolutionary burst of the flowering plants occurred well *before* the extinction of the dinosaurs. Could the small mammals have caused the extinction of the dinosaurs by developing an inordinate taste for dinosaur eggs (which some scientists believe became thinner-shelled in the late Cretaceous)? Well, possibly, but the trouble with this kind of theory is that there is no way of proving it one way or the other. What makes it rather unlikely is that there were so many other changes – which could have nothing to do with the eating habits of the mammals – going on at the same time. The mammals could have learned such eating habits in the

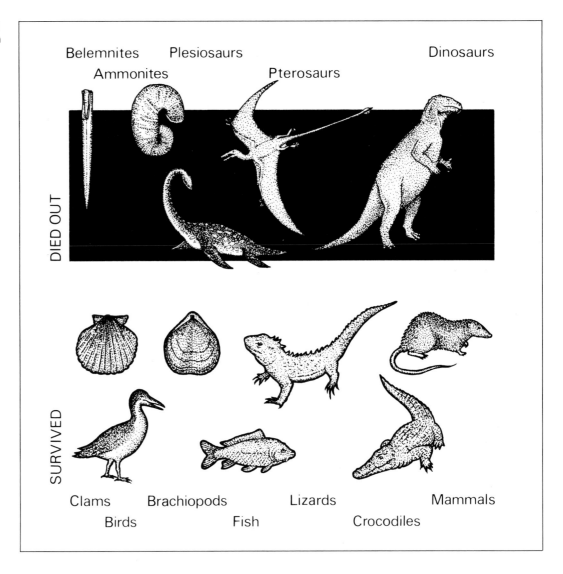

Belemnites   Plesiosaurs   Dinosaurs
Ammonites   Pterosaurs
DIED OUT
SURVIVED
Clams   Brachiopods   Lizards   Mammals
Birds   Fish   Crocodiles

mid-Jurassic, or the early Cretaceous, and it seems unlikely that coincidental changes should have occurred at the *same* time in different environments. And so it goes on: theories are proposed and defended vigorously by their adherents, only to flounder against some inconvenient fact, or belong to the category of 'pseudo-science' for which an explanation, no matter how dramatic, is not capable of being tested against the facts.

So let us return to the facts that are available about the late Cretaceous, gleaned from the rocks or from our new knowledge about the geography of the time, to see if these can throw any light on the problem. In the first place it is clear that the Cretaceous–Tertiary boundary was a time of major regression of the sea – the kind of periodic withdrawal of the ocean from continental areas that we mentioned in Chapter 5. This had been preceded by a period of widespread transgression, continental flooding, in the upper part of the Cretaceous period. This was also a time when the 'supercontinent' of Pangaea was advancing into fragmentation into the various chunks that make up the present-day continents. This process was underway as far back as the Jurassic, and was to continue through the Tertiary to the present, so that the time of mass extinction was about halfway between the time when Pangaea was entire and the situation today. It may have been true that late Cretaceous was a time of particularly active sea floor spreading, as it is certainly true that it was a time of considerable mountain-building activity; and both are connected with times of transgression. The major, sudden period of regression that followed probably reflects a sudden cessation of sea floor spreading activity – perhaps in the wake of the intense mountain building – and here we have an event that affected both sea and land, and which looks potentially promising as a way of explaining biological changes that occurred in both realms. The regressive phase has resulted in a shortage of sections of rocks which span the critical interval – the critical time is often represented by a 'gap' corresponding to the withdrawal of the sea. This has led to the suggestion that the extinction events were not quite as sudden as the record suggested, that they only seemed so because of the imperfection and incompleteness of the fossil record across the critical interval. However, now that cores taken from the deep sea have been investigated, from sites where deposition was certainly continuous across the crisis, it is found that the faunal changes appear suddenly there too, so that the event *must* have been a sudden one.

When the regression was at its height (or rather depth!) the seas withdrew from what had been diverse and rich continental seas, with many species adapted to particular ecological niches. The restriction of shallow-water areas would simply have crowded out certain of these species, but this still would not be sufficient to explain the total removal of whole groups. But there seems to have been a second effect at the time of the regression. Although the species numbers as a whole declined, there were tremendous 'blooms' of particular species of minute, marine algae at the critical time at the Cretaceous-Tertiary boundary. Plankton blooms of this kind are known to have lethal effects in present seas by consuming both oxygen and nutrients to such an extent that other animals suffocate or are actively poisoned. If this process continued for some time, then it would not be fanciful to suggest that the effect on marine species, and particularly those living in open ocean environments (including some of the ammonites and planktonic foraminifera), would have been profound. Reduction in the numbers of fish might have been sufficient to reduce the population of plesiosaurs, which ate them, to a point where they could not recover. The algal blooms themselves may have been caused by changes in the water circulation in the oceans, which are now the subject of much speculation.

But how would the events in the sea relate to those on the land? At the time of the regression one might visualize vast, flat plains exposed above the sea which might seem ideal for wandering herds of giant dinosaurs. Explanations that attempt to relate this to their extinction, which suggest in effect that the uniform habitat produced a genetically uniform population with little capacity for generating new species, seem rather unsatisfactory for many reasons. There is no real evidence that the latest dinosaurs were this widespread, for example, and with the fragmentation of the continents there would have been opportunities enough to produce non-uniformity in the dinosaur populations. One idea that might be explored is that the dinosaurs were terminated not by the regression but by the invasion of the sea which happened immediately *before*. Because many of the last dinosaurs were massive animals they would have required a lot of feeding, even if they were cold-blooded. As the seas invaded, their living area might have been reduced to the point where there was just not enough vegetation to support a viable population. The sea would also have flooded the swamps in which some of the later duck-billed dinosaurs may have lived. With their food source gone, the larger predators would also have perished. On the other hand small mammals, which were largely insect eaters, could simply have retired to the uplands, along with snakes and lizards; and crocodiles would have retreated into the streams. If this theory were correct then the last of the dinosaurs would have predated the major changes in the marine faunas by a short time, which is at least a testable idea. But as usual there are objections: why, for example should the airborne pterosaurs perish, while the birds survived and prospered? In the sea the few groups of animals that were affected least by the Cretaceous–Tertiary trauma were nearly all bottom-living organisms, gastropods, crabs and clams, that could survive adequately in the marginal seas while the regression was at its maximum.

No account of the extinctions at the end of the Cretaceous would be complete without describing the latest and most persuasive of the catastrophe theories: that the extinctions were the result of the collision of an asteroid (or several large meteorites) with the Earth at this time. When the first edition of this book was written this idea had just achieved wide currency. A great deal more information has been added since them. The original evidence, and much more that has come to light subsequently, related to an unusually high content of the element iridium, which

A dramatic representation of a meteorite storm in a nineteenth century woodcut

had been detected in a 'boundary clay' forming the junction between the end of the Cretaceous and the beginning of the Tertiary. This high 'iridium anomaly' was supposed to be the product of a major extraterrestrial impact – which would be predicted to produce such an effect. Geochemists have now investigated very many rock sections spanning the critical interval from widely separated parts of the world. In virtually all of them the same iridium anomaly has been discovered at the same horizon. Furthermore other special signs of meteoritic impact – such as 'shocked

quartz' – have turned up in these sections at the same level. In spite of the criticism that iridium anomalies can result from causes other than meteoritic impact, it would take a very sceptical mind to discount this mass of evidence. It really does seem that there was a major impact of an extraterrestrial body close to the time when the dinosaurs and ammonites died out, and many other animal groups suffered trauma. In some sections soot horizons have been detected at the appropriate level – just what you would expect if there were a major conflagration. And then afterwards there is an increase in fern pollen, which, because ferns are the first plants to recolonize after such major disasters, is also what might be anticipated.

There are a variety of scenarios which describe how the impact of an extraterrestrial body is supposed to have caused extinction. Most of them include variants on the 'nuclear winter' – an explanation which had been developed in anticipation of the catastrophic effects of a nuclear war. Vast quantities of dust are thrown up into the atmosphere, and smoke pours from massive forest fires, blotting out the sun, and this kills the vegetation which needs sunlight to photosynthesize. There is a similar catastrophic effect on the plankton in sea, which forms the base of all marine food chains. Because the largest dinosaurs were vegetarians they could cope neither with the loss of their food, nor with the cold weather. The carnivorous dinosaurs that preyed upon the grazers would soon die out when their prey was no longer available. A similar chain reaction could be envisaged in the sea. The effects might persist for several years. Survival might be a matter of being able to tolerate low temperatures, or being able to enter a state of dormancy (as a seed, or a burrowing larva, for example), or perhaps just having the ability to eat a variety of foodstuffs. Some organisms were 'pre-adapted' to survive. The small, warm-blooded mammals might have survived on a diet of scavenging insects. In the sea, crabs could have picked from the varied larder they still enjoy today.

This theory has a simplicity which makes it very attractive, if one can apply such a word to one of the greatest disasters in the history of life. It seems almost parsimonious to complain that there are some difficulties with it. The extraterrestrial impact seems very likely – but did it really cause the

The famous meteorite crater in Arizona – could exceptional impacts have been implicated in the demise of the dinosaurs?

extinctions? In the first place it is claimed that the animal groups that became extinct – ammonites and dinosaurs – were already declining before the end of the Cretaceous. If this were so it could scarcely have been in anticipation of an extraterrestrial impact. Second, it is hard to explain why some of the animals and plants that survived were affected so little. Insects and flowering plants include good examples. Cretaceous flowering plants are closely related to living ones, and most of the insects can be placed in living families, or even genera. Yet surely Cretaceous events would have affected the flowering plants as much as any group of organisms. One moth from the Cretaceous has living relatives that eat pollen, and there is no reason to suppose that its Cretaceous relative did otherwise. How could this moth have survived the darkened years? Similarly, colonial corals in the sea are exceedingly sensitive to light and pollution, yet they were not extinguished either. Already it seems necessary to introduce the notion of a corner of the world that was not affected by the catastrophe to the same degree: possible, of course, but more complicated. There are even claims that the marine extinctions may not be as profound as has been claimed, being more the product of the way palaeontologists have applied their names than a reflection of a real phenomenon.

This short account of the Cretaceous–Tertiary extinctions will show just how complicated it can be to work out the causes of one of the most important events in the history of the biological world. It is obviously worthwhile finding an answer, sifting through all the contradictory evidence to find the real causes, even if the 'right' answer always seems to elude us. The whole story could be repeated for the Permian–Triassic mass extinction. Again there was certainly a period of marine regression, but this was the time when the separate fragments of Pangaea finally collided to form a supercontinent. This aggregation was certainly connected with a change to an extremely arid environment in many parts of the world. It is possible to combine these effects together in various ways to explain why so many animals became extinct at about the same time, although the late Permian event seems to have been more leisurely than its late Cretaceous counterpart. Whatever the final explanations of the mass extinctions turn out to be, it is undeniable that their effects have been the most important thresholds through which the fauna and flora passed to make the modern world.

CHAPTER SEVEN

ORIGIN OF LIFE AND ITS EARLY HISTORY

When the earth formed about 4500 million years ago there was no life. By about 3500 million years ago the first, tentative traces of life are to be found in the rocks. So the origin of life can be placed within a broad time period. And there is no shortage of time. After all, the profound series of changes that have occurred since the Cambrian, and which are the subject of the rest of this book, took only a fraction of the whole history of the earth. But the time of origin of life is also the least accessible to direct study, all certainties about the nature of the atmosphere and configuration of land and sea are gone, and speculation has more room for manoeuvre with this palaeontological mystery than with any other. What is certain is that the earth's surface at the time of the origin of life must have been different in almost every respect from that of the present; at least some of the early conditions might be simulated in the laboratory. There is no suggestion yet of being able to create life, but it is possible to create the building blocks from which the edifice of life may have been constructed.

The formation of the earth was but one incident in the formation of the solar system as a whole. If present ideas are correct, the earth *accreted* from small particles swirling in a disc-shaped nebular cloud. The planet grew from a 'seed' of magnetic metallic grains, which attracted more material by gravity, so that it grew rather like a snowball. Early on, the Earth melted and a molten core was formed, largely composed of the metals iron and nickel. Accretion of other material continued, with the addition of volatile components like water, carbon dioxide and chlorine at a late stage. It is unlikely that the earth retained much of an atmosphere at this time. Then the surface of the earth quickly cooled to form a thin brittle crust. Bombardment with meteorites continued; the surface of the moon preserves, as it were frozen in time, this stage in the evolution of the earth. But the earth, unlike the moon which cooled quickly, incorporated the additional meteoritic material into the upper part of its mass. The gravity of the earth was powerful enough to 'hold on' to some of the more volatile ingredients, particularly water, without which there would have been no life. The atmosphere gradually formed around the earth, and the incessant bombardment by meteorites slowly abated. The rate of heat flow from the earth's interior was high at this early stage, and volcanic activity was probably widespread and continuous. Any volcanic eruption is accompanied by the release of huge quantities of steam and gases like nitrogen and carbon monoxide as well as highly toxic acids like hydrogen chloride and sulphur dioxide. We can visualize acid rains attacking rocks and reacting with them to form such widespread minerals as sea salt (sodium chloride). The steam released into the atmosphere from the volcanoes condensed to form seas, and since the first traces of sedimentary rocks are as old as 3800 million years or more, there was water enough for sedimentation by this time. And presumably also the processes of erosion had begun to shape out patterns of cliffs and shores so that the first vestiges of the modern topography had appeared.

But in several important respects the environment was completely different then. There was little or no free oxygen in the atmosphere. This meant that there was no 'ozone layer' – a cloak of modified oxygen high in the atmosphere today that acts as a screen to ward off ultraviolet and other harmful radiation from the earth's surface. The penetration of such light to the surface produced the kind of chemical reactions that might have led to the necessary building blocks for life itself. The water molecule may well have been 'split' into its component atoms this way too. These chemical reactions can be reproduced in the laboratory – the right mixture of chemicals and gases, charged with the effects of ultra-violet light – and sure enough some of the basic ingredients of life appear after a while.

The most recent ideas are that the original atmosphere may have had a large proportion of carbon dioxide. The presence of *amino acids* is particularly important, for chains of such acid molecules, joined together in their hundreds to produce giant molecules, are the basic structure of proteins, without which there is no life. Some protein-like chemicals have even been synthesized in this way, and in a brine medium tend to coagulate into tiny spheres, which have at least a passing similarity to living cells. They are in no sense alive, however, because life demands the ability to reproduce. All organisms feed (or, if they are plants, manufacture their own food), and are able to produce copies of themselves. Presumably it was necessary to have evolved the long spiral chain molecule DNA, or a part of it, which is the basis of the copying process, before the first self-replicating cells could exist. So there is still a big gap between the sort of molecules that were cooked up in the 'primeval soup' and the first living cells, which were probably of the simplest kind, like bacteria. Some primitive living bacteria have an 'economy' built on exploiting sulphur compounds, which were present in abundance around the hot springs, fumaroles and volcanoes in the remote Precambrian. Whether the vital steps occurred in the early seas or in some other site, such as shallow pools on the hot slopes of volcanoes, is still a matter of debate. But presumably once the first true, self-replicating cell had appeared it could spread and prosper unhindered wherever the right conditions for its nourishment were to be found. That earth, no matter how inhospitable it would seem to us, would have been a Garden of Eden for the first organism. Most living organisms share so many molecular features in common that it seems that the generation of life happened once only, although there is a case for

regarding some of the bizarre sulphur bacteria as the earliest side branch (or perhaps an independent line) on the complex chain from bacterium to mammal.

This explanation of the origin of life requires nothing but time, the right conditions, and a few special events to link all the components into the finished cell. Recently another idea has claimed a lot of attention, although it is not a new one. Life may have been of extraterrestrial origin. The early earth was almost certainly bombarded with particular kinds of meteorites (known as carbonaceous chondrites) and these meteorites when they appear at the present time *do* contain organic kinds of molecules.

As far as the origin of life on earth is concerned it is true that it is possible to produce carbon compounds like those in meteorites under the right experimental conditions. But there is a curious difference between these compounds and those which predominate on earth. Many carbon compounds can exist in two forms, which are chemically identical, but which have the property of rotating light either to the left or to the right. On earth there is a predominance of the former which rotate light to the left; but from meteorites the left and right-handed forms of the same compounds are about equally abundant. So it begins to look as if organisms on earth manufacture carbon compounds which are subtly different from those produced in meteorites, or under laboratory conditions to simulate meteorites. All in all it is probably preferable to regard the earth as the cradle of life, if only because this theory seems to require fewer co-incidences to produce an organism adapted to terrestrial conditions, and there was a vast stretch of time available for the shuffling and rearrangements of chemicals necessary to make the first living cell.

Of course, although meteorites contributed organic molecules to the 'soup' from which life was brewed, it is likely that the cauldron itself was on the earth and not in the far reaches of space.

Some scientists have claimed that the periodic appearance of extraterrestrial materials on earth, especially from comets, has continually influenced the evolution of life – not just its earliest history. For example, they envision such processes introducing alien biological material as an explanation for catastrophic viral epidemics. However, meteorite induced epidemics are another of those theories about extinction which it is impossible to test one way or the other; in any case it is difficult to see how they could have been so selective as to remove, say, dinosaurs but not birds (which may be close relatives of the dinosaurs), or ammonites but not snails at the end of the Cretaceous. Comets passing close to the earth from time to time may well be implicated in some of the crises in the history of life, but this is more likely to have been by secondary effects rather than by direct biological influence. Viruses are rather specialized organisms, probably secondarily simplified as much as genuinely primitive, and often adapted to a particular host. A generalized virus with fatal properties arriving from 'out there' might be expected to have removed *everything*, or if somehow preadapted to one particular organism to have removed *only* that species. Neither circumstance would explain mass extinctions as we actually observe them in the rocks.

One way or another the crucial steps were taken, and cells evolved. It is only in the last decade or two that proof of the existence of the oldest kinds of organisms has been recovered from the rocks. As one might expect the rocks are rather special

ones: they have survived from the remotest parts of the Precambrian hardly altered by heat or pressure, which is an exceptional circumstance in itself. They are often *cherts*: fine-grained rocks composed of silica, very hard, and when cut into thin slices and ground away they become transparent. It was only when this was done that the tiny fossils were revealed. Most of the cherts which have yielded fossils seem to have been deposited in fairly shallow water around the protocontinents of the time.

The earliest and most primitive fossils are those of *prokaryotes*, simple rods, spheres or filaments in which the cell contents lack a defined nucleus – the 'package' within the cell which houses many of the vital organelles of all higher organisms. Such simple prokaryotes are known as fossils in rocks as old as 3500 million years (see below). Fossils of *eukaryotes* with a defined nucleus are known from rocks about 1700 million years old. Even for those used to thinking in terms of geological time these figures are hard to comprehend. The time during which these simplest of organisms held sway is so much longer than that which has elapsed since the beginning of the Cambrian – which used to be thought of as comprising the whole of the fossil record.

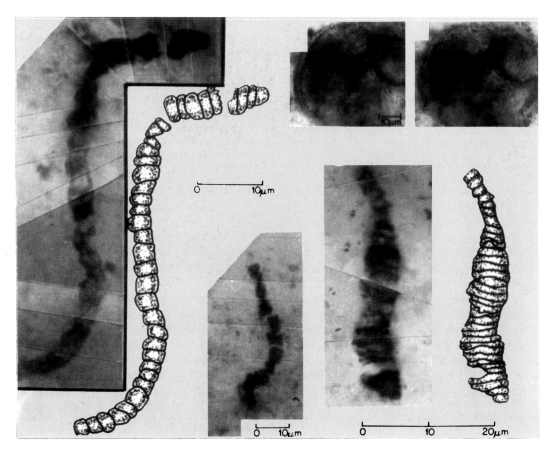

The oldest fossils. Minute threads and spheres 3500 million years old photographed from thin sections in chert from the Precambrian of Australia

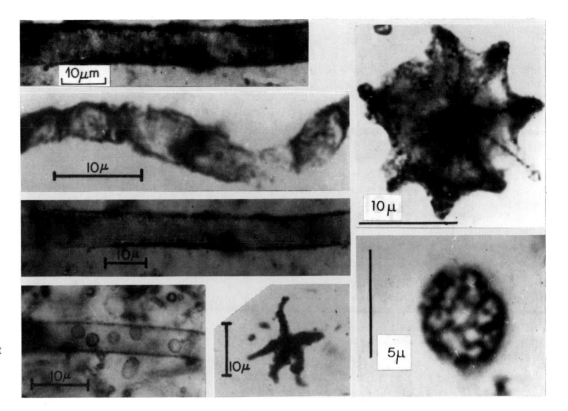

Ancient fossils from the Gunflint Chert of Canada, about 2000 million years old, photographed in thin section

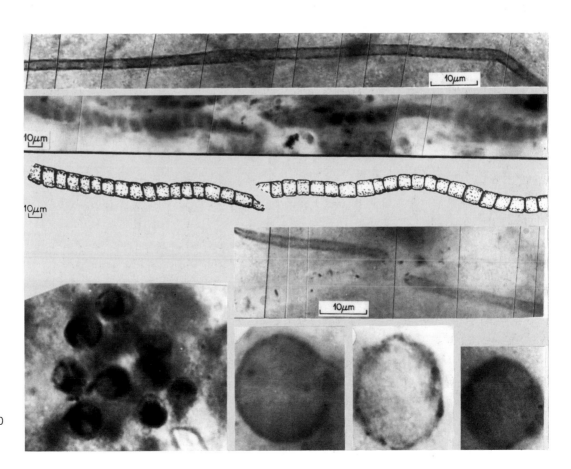

Recent discoveries from the Precambrian Gaoyuzhuang Formation of China (about 1450 million years old) include a number of prokaryotes

Momentous events at the cellular level happened during this vast time. The fossils discovered from cherts can scarcely represent the whole story, but new discoveries are still being made, and it is clear that there was an increase in the variety of these small fossils through the Precambrian, even from the patchy record they have left.

Ancient bacteria were able to live in conditions which would be inimical to most living organisms. Some had a metabolism based on sulphur, and these survive today in hot springs and similar habitats. Others acquired the ability to use the sun's energy to produce nutrients (the process of *photosynthesis*), but not all of these also gave off oxygen, as do the higher plants. Cyanobacteria which *did* exhale oxygen were certainly present at 2700 million years ago, and some scientists claim that they may have been present already at 3500 million years ago. The boundary between algae and bacteria is not an easy one to draw among these ancient and simple organisms, and different authorities disagree on the affinities of certain fossils, though all agree that these are among the most important fossils we know.

Some Precambrian fossils had been known for many years before it was settled to everyone's satisfaction that they really were organic. These include finely layered rocks, usually limey, but often replaced by chert, sometimes with the external appearance of numbers of small cushions, or occasionally steeper pillars. It was the regularity of their layered structure that suggested they might be of organic origin: such fossils were christened with names like *Cryptozoon* ('hidden animal') when they were discovered in Precambrian rocks from Canada and elsewhere. At the time it was considered rather unlikely that fossils could occur in rocks as old as this. Nowadays, fossils of this kind are known to be the remains of structures produced by cyanobacteria, a perfectly reliable indication of biological activity. They are known as *stromatolites* (*see* p. 145). Structures resembling stromatolites are now known from rocks younger than Precambrian as well; indeed, they can be rather common in limestone rocks originally deposited in shallow-water environments in the tropics, and they appear to be especially characteristic of what were quiet water sites between high and low tides. A few years ago *living* stromatolite mounds were discovered in Shark Bay, Western Australia,

and like many of their fossil counterparts they were found mainly in an intertidal environment. Since the oldest of the fossil stromatolites were about 3000 million years old, these rather unspectacular mounds take all prizes for the most enduring of living fossils. Why they had not been recognized as of organic origin is because the very primitive organisms that make the mounds are not themselves usually preserved as fossils – their simple threads generally decay without trace. The mounds are produced by a skin of cyanobacteria trapping sediment season after season to produce the fine laying characteristic of the stromatolite.

Late Precambrian Bitter Springs Chert (about 850 million years old) has yielded a great variety of microorganisms, including the spiral *Heliconema*, and cells apparently in the process of division, and definite eukaryotic organization

Fossil stromatolite cushions from the Precambrian rocks of eastern Siberia. Half size.

Beneath the surface layer other, ancient kinds of bacteria thrive, so that the stromatolite is really a very simple community, one which has survived from the earliest days of the Earth. Tidal waters drain off through the channels between the mounds. In the living examples (since Shark Bay additional sites have been discovered where stromatolites are forming today), there is a certain variation in the form of the mounds according to where they are growing. Some of the same variation has been observed in the fossil forms. Nonetheless, a growing number of palaeontologists studying the Precambrian mounds believe that there is also a variation through time in the shapes of the mounds so that they can be used to characterize very broad segments of the Precambrian. It would now be true to say that their occurrence in Precambrian rocks which were deposited over the epicontinental seas of that time is almost ubiquitous; they have been found over huge areas of the USSR and Australia, Greenland, Africa, Canada, the United States and even Scotland. Obviously they were much more widespread in these far-off times than they are today. This is probably because there were no Precambrian animals adapted for grazing, one of the standard ecological niches in living marine faunas. Living stromatolites owe their existence to special conditions inhibiting such grazers. So it is not altogether fanciful to visualize vast flats covered with stromatolites, enduring for a period of more than 2000 million years of the Precambrian.

Cyanobacteria and algae produced free oxygen during the process of photosynthesis. This process was probably the most important ingredient for setting up the conditions for the evolution of all higher organisms. The photosynthetic activities of cyanobacteria and algae could

Living stromatolites: algal mounds forming in Shark's Bay, western Australia

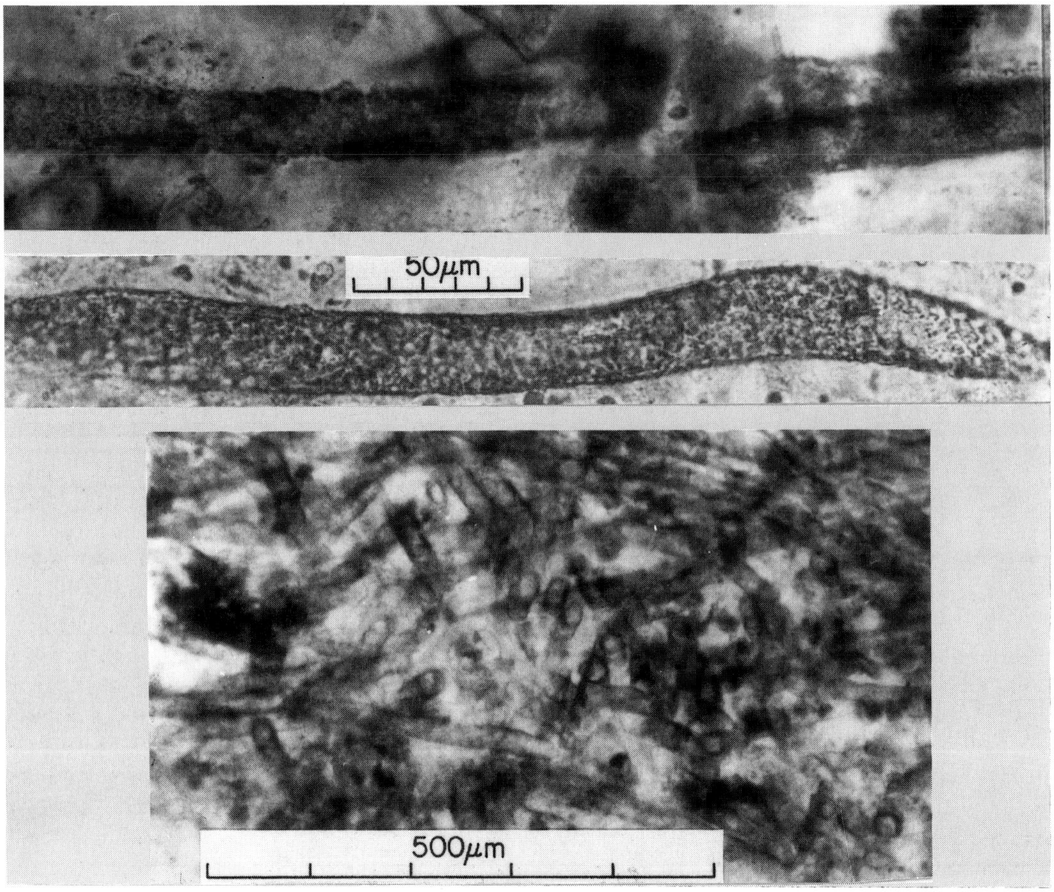

50μm

500μm

proceed uninterrupted for more than 2000 million years. If they were as widespread as seems likely they were capable of adding oxygen to the atmosphere. Little by little an atmosphere was created in which other organisms could breathe; they made the environment fit for animals. It is possible to trace the gradual enrichment of the atmosphere in oxygen by changes in the types of sedimentary rocks that were laid down during the Precambrian. In the earlier part of the Precambrian when oxygen-poor (reducing) conditions prevailed, there are widespread deposits of a distinctive kind of banded iron ore, which could only form when the atmosphere was low in oxygen. These banded iron ores are almost never found in rocks younger than 1000 million years old; the few younger occurrences are small and isolated. In the younger rocks iron-bearing beds are usually of a brighter red colour (they are known as 'red beds' for this reason), with the iron in an oxidized state, which, not surprisingly, indicates the presence of a good deal of oxygen in the atmosphere. There must have been a critical point when there was enough oxygen in the atmosphere to permit respiration – so that it was possible at last for animals to exist that

needed to breathe oxygen, and perhaps begin feeding on the plants that had made their existence possible. At the same time the atmosphere would have acquired its protective 'ozone layer' which would cut out some of the more harmful effects of solar radiation.

Cherty rocks from the Precambrian have now yielded many remains of cyanobacteria and other bacteria, some of which have been discovered since the first edition of this book was written. A variety are illustrated here. Most of the early species reproduced by fission, that is, their cells simply divided into two, so that the daughter cells were identical to the parent cell. Not surprisingly this mode of reproduction leads to large numbers of cells very quickly but their capacity for change is limited. Any spontaneous change (mutation) in the genetic make-up of the cell is likely to be disadvantageous, even fatal. To use an analogy, imagine the chance that a conveyor belt in a factory would turn out a better product if the machine manufacturing the goods developed a fault. In spite of this there is a lot of evidence from experiments with living bacteria, for example, that advantageous changes can and do happen, and that when such a change occurs its spread into a whole population is extremely rapid.

From the later (850 million years b.p.) Precambrian Bitter Springs Chert of Australia more complex cellular arrangements have been found (see p. 144). Because these fossils preserve even some of the details of the inside of the cells their preservation must have been extremely rapid. We have to visualize the cells being impregnated with silica as they were still alive, in something like the way the Devonian Rhynie Chert preserved the fine details of much later plants (*see* Chapter 1). Cherts with ages of about 1000 million years have yielded the remains of cells which seem to be undergoing cell division of a different kind from simple fission. In these cells there is evidence of the *combination* of two cells from different plants of the same species, with subsequent division, and sorting out of the genetic material from the two parents. Some miraculously preserved specimens even *seem* to show the nuclei that carry the genetic material in the process of undergoing this kind of division. These fossils are particularly important because they show that plants had evolved different sexes at this time. Cross-breeding of the kind that had now evolved had several important results, but particularly it increased the chances for the inheritance of advantageous changes. It must have been a stimulus for further evolution.

There is evidence that the open oceans were colonized at a similar time. From about 1000 million years onwards some of the more oceanic sediments that were deposited in the Precambrian have yielded the tiny cases of single-celled algae called *acritarchs*. If the oceans then had something like the dimensions they do today it would be difficult to overestimate the importance of these tiny plants, for their photosynthetic activity would have greatly boosted the quantity of atmospheric oxygen. They were also the foundation of a food chain. Even today most of the animals that live in the ocean depend ultimately on minute photosynthetic algae, which provide food for the minute planktonic animals, which are food for larger fish, which in turn feed sharks, and so on. Some scientists believe that the Precambrian ancestors of most of the major animal groups were small planktonic animals, just as their larvae are today. But, alas, there is virtually no fossil evidence of these tiny animals, if they were indeed feeding on single-celled algae in the seas of the later Precambrian. There is no doubt that they would be extremely difficult animals to find in the fossil state, because of their lack of hard parts and small size, and we can only hope that somewhere they will be preserved by the kind of geological miracle that allowed the discovery of the early plants and bacteria.

In popular accounts written only a few years ago (and regrettably in some only just published) it is usual to find the Precambrian described as 'lacking fossils'. This is obviously untrue. Of course, many Precambrian rocks *are* devoid of organic remains because they have been metamorphosed, if they were originally sedimentary. But it is surprising in how many of the rock sequences that have escaped metamorphism some organic traces have been discovered. So far the emphasis has been on the bacteria and algae which served to prepare the world for animals. It is now usual to refer to the later part of the Precambrian as the *Proterozoic* era, with a status equal to that of the Palaeozoic which follows. No doubt as more people look for fossils in these ancient rocks more will be discovered.

If the earliest history of the animal phyla is still obscure, there are now quite a large number of good animal fossils from the later part of the Precambrian, and more seem to be turning up each year. It is still true to say that almost none of these Precambrian animals had hard parts – shells, bones or the like – and so the circumstances in which they have to be preserved are somewhat special. On the other hand, conditions in the late Precambrian were different from those that followed in the Palaeozoic, and some of these conditions may have favoured the preservation of soft-bodied animals. For example, it is quite likely that scavenging animals had not evolved at that time, so that the bodies of dead animals were permitted to 'lie around' longer than they would at the present day, so increasing their chance of preservation. Many of the early animal fossils are of jellyfish (*see* p. 148). Some of their stomach cavities are preserved as casts in fine sediment. A few of these jellyfish may even be related to forms still living, and it is reasonable to suppose that they had similar habits, drifting about in the open oceans carried by currents, feeding on other planktonic organisms. Their stratigraphic position below the base of the Cambrian has been known for some

time; *Brooksella*, a jellyfish from the old rocks at the base of the Grand Canyon sequence was discovered fifty years ago, while some of the Canadian examples were discovered almost a century ago. It was some time, however, before they were accepted as 'real' fossils. Even now, some of the supposed Precambrian jellyfish have been shown to be only a rather peculiar kind of sedimentary structure, entirely inorganic, produced by the expulsion of water from wet sediments. But more and more examples have now been found, many of then indubitably true fossils, and their worldwide distribution is impressive: several localities in Canada and the United States, Australia, many finds in Russia, with scattered occurrences in Europe, including Britain. A new discovery of Precambrian jellyfish was made in Wales as recently as 1978. We might anticipate that a free floating organism like a jellyfish would have a wide geographical spread, but it is surprising to find that so many localities had the right sedimentary conditions for their preservation. It is clear that the late Pre-

cambrian oceans supported a variety of planktonic single-celled algae, and many drifting jellyfish; almost certainly there was also a variety of small, soft bodied planktonic animals, perhaps resembling the larvae of living marine organisms, which provided food for the jellyfish, and which may also have included ancestors of many of the groups of animals that appear 'ready made' at the base of the Cambrian.

However, there are also a few late Precambrian fossil faunas that include a much wider variety of fossils than simple jellyfish. The most famous of these is probably the Ediacara fauna from southern Australia, where there is an exceptional preservation of many soft-bodied organisms. The true affinities of these impressions is still under debate. It is possible that segmented worms (annelids) are represented by such fossils as *Dickinsonia* (*see* right). Some other curious segmented animals have been suggested as arthropods, but this is far from generally accepted. If it is true that the arthropods were derived from a segmented, worm-like animal, then we might expect to find the

common ancestor of all the arthropods in the later part of the Precambrian period. The peculiar animal *Tribrachidium*, with its three curved grooves, has been suggested as an ancestor of the echinoderms, but there is really not much evidence to support this. A common element in the Ediacara fauna, and one which has turned up very widely from other parts of the world, including Newfoundland and Charnwood Forest, England, are frond-like fossils which look rather like the living sea pens. Some of these were attached to disc-like bases. Even these animals have attracted a lot of controversy about their true affinities. If there was ever any idea that by finding soft-bodied fossils in the late Precambrian it would be a simple matter to relate these to the living animal phyla, the Ediacara fauna has shown that the real world, as usual, is more complicated. Instead there seems to have been a distinctive late Precambrian fauna with a number of curious animals all of its own. This impression is confirmed by the discovery of other faunas in the USSR and

Newfoundland which include some forms in common with the Ediacara, but have others unique to them. The Newfoundland fauna occurs in a series of beds exposed at Mistaken Point, a promontory on the south of the Island. Here centuries of weathering have etched out the impressions left by the soft-bodied animals, which are left like a picture gallery of the distant past, now washed by the waters of the present Atlantic Ocean. The age of the Ediacara fauna is not much older than the base of the Cambrian, but the Newfoundland faunas may be rather older. So the peculiar animals were evidently widespread, although rarely preserved as fossils, and may have held sway over the late Precambrian sea for perhaps a hundred million years.

Apart from the jellyfish, these animals were probably bottom dwellers. Even in rocks where the remains of soft bodied animals are *not* preserved you might expect to find the traces left behind by the activity of animals in the sediment – burrows and trails and the like. Oddly enough, trace fossils are relatively rare in Precambrian rocks in general, even though they are quite common in rocks of early Cambrian age from many parts of the world. A number of supposed trails have been found in rocks going quite far back in the Precambrian (in excess of 1000 million years), and some of these may be genuine, but the trouble with the more generalized kind of trace fossil, a wandering squiggle on the surface of a lime-stone perhaps, is that similar-looking marks can also be produced by inorganic causes and so whether you interpret them as animal or not depends on how much you want to believe that there were animals around at the time. In some sections of rock which pass upwards continuously between the Precambrian and the Cambrian, in Arctic Norway for example, it is noticeable that the numbers and variety of trace fossils seem to increase as the boundary with the Cambrian is approached. Once the boundary is passed it is easy to find deep burrows, vertical to the sediment surface. Perhaps it is true that deep burrowing habits were only acquired by animals of Cambrian and younger age.

No survey of the early history of life would be complete without taking the story over into the lowest part of the Cambrian period. The old idea that life appeared quite suddenly at this boundary will by now be seen to be entirely wrong; we have traced the gradual increase in variety of organic

Some of the oldest fossils with hard parts that appear near the base of the Cambrian; few of them can be easily related to living organisms. They are all a few mm across

150

remains through the Precambrian from about 3500 million years ago. The evidence is still tantalisingly incomplete and it is still possible that one discovery which miraculously preserves a whole fauna and flora could change the story in many details. The last few years have also seen a great increase in the effort put into studying fossils from the earliest Cambrian rocks. Although the Cambrian rocks were first distinguished from Wales, the earliest beds of the system are not well-developed there, and the most important rock sections for studying this interval of time are from other parts of Europe, China, Morocco, Australia, Canada, the United States and Siberia. These are mostly sediments deposited in relatively shallow-water marine environments. The best approach to understanding what happens in the very earliest Cambrian is to collect fossils bed by bed through the critical interval, charting the first appearance of the various kinds of animals the rocks contain.

It is a remarkable fact that animals with hard parts appeared widely at the base of the Cambrian. For several thousands of millions of years there had been life, but no organisms had secreted a shell, or any kind of preservable strut to serve as a skeleton. Within what must have been a comparatively short time, many different kinds of organisms had apparently been able to do just this. It is still something of a problem to be sure from the appearance of these hard remains from so many different sections in different parts of the world was at *exactly* the same time. But it is fairly certain that it happened within a very short time period, using the fossils themselves as the means of correlating the different sections. Not all of the various kinds of animals that are characteristic of the Cambrian made their appearance together in the earliest rocks. There was a time lapse between the very first hard parts to appear in the rocks, and the appearance of some of the animals which are familiar in later Cambrian and Ordovician rocks. There are no brachiopods among the earliest faunas, for example. And many of the organisms that appear first are very peculiar animals indeed. Many of them are geological enigmas, like *Janospira* (p. 122) from later rocks. They simply cannot be fitted conveniently into the groups of animals we know about from younger rocks. A lot of them are rather simple tubes, sometimes twisted, and with differing cross-sections.

They could have been secreted by any number of different 'worms', or by something else for that matter (*see* p. 122). Then there are a number of odd, cap-shaped shells, which are, if anything, even more puzzling. Some of them appear to fit together in a kind of mosaic, and very few of them are obviously related to anything else that follows. Yet some of them are also very widespread, so whatever kind of animals produced them, they were evidently fairly successful for a short time. In some sections these peculiar assemblages of fossils precede the rocks that contain more familiar kinds of animals, like trilobites. Perhaps these early oddities were 'experiments' that had a short heyday, but did not survive long once the typical Cambrian faunas were established.

Among the early fossils were the peculiar archaeocyathids, a distinctive group of sponge-like forms which produced the earliest structures that might be called reefs. These also made their appearance among the earliest faunas with hard parts, and did not long survive the Lower Cambrian. The kinds of limestone that contain archaeocyathids also yield remains of calcareous algae – that have secreted a calcium carbonate skeleton around their delicate threads – so it was not only animals that acquired the ability to build skeletons at this time. Trilobites regularly appear among the earliest, but not generally the very oldest, Cambrian faunas, and often there a number of different species together, which must indicate that they had a history of which we know nothing in the latest Precambrian. The earliest trilobites are already quite complex animals (*see* p. 153), with well-developed eyes. The earliest molluscs are also present in early Lower Cambrian rocks, and include representatives of the gastropods, chitons and the monoplacophora (p. 77) but many groups of molluscs are hard to trace below the Ordovician. The earliest echinoderms from the Lower Cambrian include some very odd animals that are not easy to relate to any of the echinoderms that are prominent in recent oceans, most of which have histories going back to the Ordovician. The novelties include the extraordinary helicoplacoids (p. 84), which, like so many of the more peculiar animals, did not survive the early Cambrian faunas in general, and lower Cambrian ones in particular. Many of the animal groups which dominate Ordovician and younger rocks have only the sketchiest

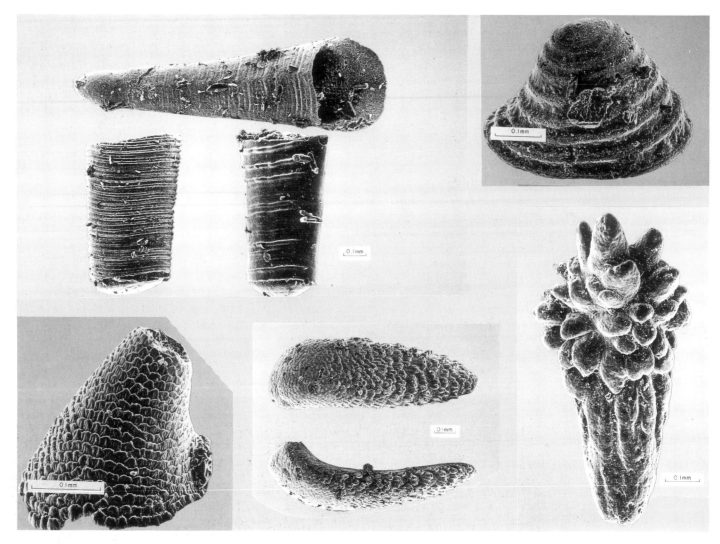

Some of the early Cambrian small shelly fossils from South Australia – with disputed zoological affinities

Cambrian histories – admittedly we can recognize other representatives of the same phylum, but these tend to be rather odd animals, with a history confined to the Cambrian.

There are further peculiarities of these early faunas. There are a surprising number of early Cambrian animals, including many of the odd ones, that seem to have used calcium phosphate as the material to build their skeletons. Among the animals that dominate Ordovician seas right through to the present, calcium carbonate (calcite, aragonite) is by far the dominant building material for skeletons (with the important exception of the vertebrates). Two kinds of animals used silica for the construction of their skeletons – some of the sponges and radiolarians – and these two groups also have a history extending back into the Cambrian. When skeletons were first produced, it looks as if there were no particular advantages in using one material or another, but subsequently calcium carbonate proved the more generally adaptable. Whether there were any conditions that particularly *favoured* the use of calcium phosphate in early Cambrian times is more difficult to say, although explanations have been put forward which related phosphate deposition to the gradual changes that were happening in the seas and atmosphere, as conditions became more like those pertaining today. In any case, the common occurrence of phosphatic skeletons is yet another reason to regard the early Cambrian faunas as distinctive.

So we come to the critical question. Why did the appearance of hard parts apparently happen within such a short time? In the past, before so much was known about Precambrian animals, the answer might have been that there *were* no animals until shortly before the Cambrian. The sudden appearance of the various animal phyla would then have naturally been accompanied by the sudden appearance of their shells. Today, there are advocates of such a Cambrian 'evolutionary explosion' who maintain that the faunas of Ediacara type fell victim to a major extinction, and that the Cambrian appearances were the result of a subsequent period of frantic and accelerated evolutionary change. The alternative view would emphasize the fact that even in the earliest part of the Cambrian the types of organisms with shells were already highly varied showing that evolution had been proceeding on a number of independent lines for a long time. Therefore, the reason for the ability to secrete a shell must lie with some factor outside the animals themselves. It would be hard to believe that such independent lineages should by chance alone 'decide' to place a premium on acquiring hard parts at the same time. It may be possible to fit the Cambrian events into the broader picture of the evolving atmosphere, and the changes in the earth as a whole, which I have sketched out earlier in this chapter. There is some evidence to indicate that it was not possible for calcium carbonate (or perhaps the other skeletal minerals) to be deposited by living tissues until the pressure of oxygen in the atmosphere had reached a critical level. Possibly this level was reached near the base of the Cambrian (although many experts place this event much earlier); if conditions suitable for phosphate deposition had been reached slightly earlier, this might explain why there are so many phosphatic shells at this time.

There was also a widespread glaciation in the late Precambrian. There has been the suggestion that the sudden appearance of animals with hard parts may have been related to this event. Glaciation would have lowered sea level, and if the effects were dramatic enough this could have exterminated the peculiar animals lacking shells that are found in the Ediacara fauna and elsewhere. Subsequent elevation of sea level following the melting of the Precambrian ice caps would produce a worldwide marine flooding, and this has been repeatedly observed in many sections which span the critical interval. The base of the Cambrian is often marked by shallow-water, even intertidal, sandstones, which record the former

One of the earliest trilobites from the Cambrian: already quite a complex animal

shorelines as the sea transgressed over the eroded Precambrian landscape. The base of the Cambrian would then mark not only the appearance of skeletons, but also a re-colonization of an environment laid waste at the end of the Precambrian; and it is true most of the earliest Cambrian animals seemed to have been inhabitants of the shallower seas. Presumably this recoloniz-ation would have to come from the plank-tonic animals (or perhaps the deep sea) in areas protected from the glacial crisis. Planktonic animals and plants, including the jellyfish and single-celled algae, would have been best adapted to surviving the peak of the glaciation by retreating to the open ocean, and it is indeed true that of the Precambrian organisms these two groups alone pass into the Palaeozoic and beyond. If this explanation were correct then the earliest Cambrian would have been a period of extraordinarily rapid evolution — an 'evolutionary burst' like the radiation of mammals in the early Tertiary. Perhaps the ancestors of many of the early Cambrian forms were planktonic animals, and perhaps it is not coincidence that all major phyla still have planktonic larval stages. The growth of the shell may have been a response to settling on to the sea bottom. All this is speculative, and very difficult to test one way or the other. In some sections the glacial event appears to have been too early to have had such a seminal influence. The acquisition of shells and skeletons is one of the great milestones in the history of the biosphere, and the difficulty of finding a single neat explanation only adds to its fascination.

CHAPTER EIGHT

FOSSILS IN THE SERVICE OF MAN

Every time you drive a car you are able to do so because of the photosynthetic activity of plants many millions of years ago. The growth of industry to the level it reached in the nineteenth century was as related to the exploitation of coal as the world economy is related today to the extraction of oil from the rocks. Most of the articles that we take for granted, from plastic spoons to television sets, ultimately owe their existence to energy derived from consuming fossil fuels. In times of inflation in oil prices this dependence becomes manifest in rises in the prices of all sorts of other commodities that seem at first glance to have nothing whatsoever to do with oil. It would not be overstating the case to say that western society owes its present affluence to fossil fuels. But as each year passes the resources dwindle alarmingly, and it has become a cliché of the times that the process cannot go on indefinitely. Man is exploiting the past, plundering the fossil record, and this can only be done once.

So far we have looked at fossils as they have been used to interpret the history of the earth, and as of interest in their own right. It is appropriate now to take a look at them from the ways in which they are of practical use, directly or indirectly. Since the initial impetus which stimulated people to look more closely at fossils was economic, notably the need to produce better geological maps, there has been a constant interplay between the academic side of palaeontology and the ways palaeontological results can be used for industrial purposes. For such purposes it is neither here nor there to know how trilobites or dinosaurs lived, and even the problems of classification, which exercise

the minds of many palaeontologists, hardly touch the more practical world at all.

MICROPALAEONTOLOGY

The smaller the fossil, the more likely it is to be of use industrially. Generally speaking the abundance of fossils is inversely proportional to their size; the tiny ones are the commonest. The chances of finding an identifiable dinosaur or fish from a section through rocks from a borehole are very slim indeed, and so the kinds of animals that dominate the landscape in reconstructions of the past do not figure prominently in the records of the oil companies. The study of microscopic fossils – micropalaeontology – has become more and more important over the last twenty years. Not only can microscopic fossils be recovered from boreholes with a diameter of a few centimetres, but they can also be teased out of rocks otherwise devoid of organic remains. Some types of microfossil seem to have evolved almost as fast as their larger contemporaries, so they can be used in just the same way, as clocks to measure the passage of geological time. It is their usefulness in stratigraphy which gives them their industrial importance. To correlate between one borehole and the next, or between a whole series of boreholes and those in a different country, one of the simplest and cheapest methods is to identify the fossils. Some of the fossils that are used were introduced in Chapter 4, but some of the others need a little explanation here. Now that microfossils have even been found in the later Precambrian rocks, their use spans almost the whole of the geological column. Palaeontologists who are experts on tiny fossils are employed by geological surveys, and oil and mineral prospecting companies.

CONODONTS

Conodonts are tiny, tooth-like fossils a millimetre or so long, made of calcium phosphate. They can be abundant fossils, occurring in thousands from a kilogram of rock. They are used as important stratigraphic indicators in rocks ranging in age from Cambrian to Triassic, when the conodonts apparently died out. In life, individual conodonts occurred together in clusters consisting of opposing pairs of identical conodonts; several different kinds of conodonts often went into these 'apparatuses'. Both the assemblages and the elements making them up have been named.

Conodonts are of special use in dating limestones. Because of their phosphatic composition they do not dissolve in acetic acid, so if limestones are put into a bath of acetic acid all the calcium carbonate dissolves leaving behind a residue of insoluble products, including all the conodonts. Limestones from a borehole can be processed almost on a production line to

give a series of ages. You do not have to be able to see conodonts on the surface of a piece of limestone to be able to find them by solution. Hundred of different kinds of conodonts have now been described.

When the first edition of this book was published, I described the conodonts as a 'palaeontological puzzle' because the animal of which conodonts formed a part was unknown. They could not be placed in a phylum with any confidence. There was a great deal of speculation about what kind of animal yielded these little fossils. Several claims to have discovered 'the conodont animal' had been made, but none of them had stood up to scrutiny. It was only clear that whatever kind of animal, or animals, yielded conodonts, there must have been a great many of them to produce so many fossils.

All this has now changed. The real conodont animal has been discovered. As so often in palaeontology it happened almost by chance. Two of my colleagues, Dr Derek Briggs and Dr Euan Clarkson, were studying the Carboniferous arthropod fossils from the Scottish Carboniferous, and

The discovery of the conodont animal in this Carboniferous specimen ended decades of speculation about what kind of creature carried these tooth-like structures

Conodonts: tiny tooth-like fossils (but not teeth) as puzzling as they are useful in dating rocks. Natural size: most conodonts are 1–3 mm in length

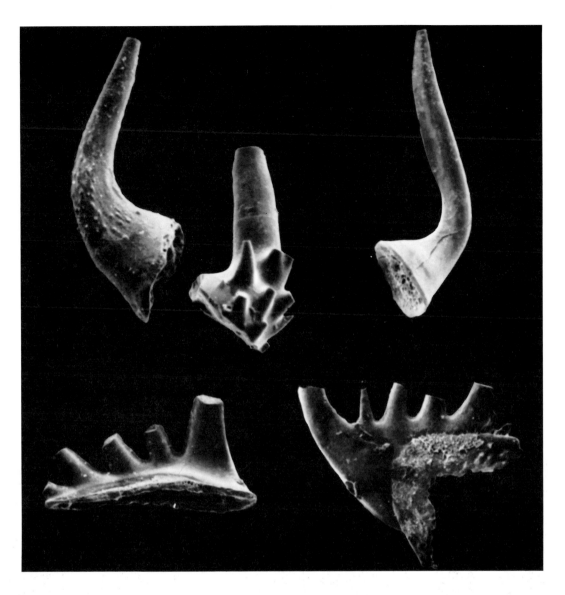

conodonts were the last thing on their minds. While going through collections in the Royal Scottish Museum a curious, worm-like fossil a few centimetres long came to their attention. It was not very conspicuous, which is presumably why it had escaped attention before. Only later was it apparent that there were conodonts within this creature. At once the conodont expert Dr Richard Aldridge was called in, and he was able to say that not only did the conodonts belong to a known variety, but they were associated together in a natural 'apparatus'. The chance of this being a fortuitous association was remote. Subsequently, more specimens have come to light, which confirms the correctness of this judgement. The conodont apparatus occupied only a fraction of the length of the whole animal (*see* p. 156). Although we now know the animal that bore conodonts this

does not solve all the problems because the animal itself was a peculiar one. The long, slim body turned out to be no worm at all, but was almost certainly a chordate – a member of that great group which includes the vertebrates. The conodont animals were a successful side branch, which apparently left no descendants. They may have been plankton feeders, and this may account for the abundance of conodonts in the rock. It is this fact which makes them so useful in geological dating – regardless of their zoological affinities.

OSTRACODS

It is quite common to find bedding planes of limestones or shales covered with tiny oval blobs looking like so many diminutive beans. These are the shells of ostracods, a group of small crustaceans which are encased in a pair of shells, looking rather like miniature clams. Like all other arthropods they have paired appendages, which they use for feeding and for moving about, and the superficial resemblance to molluscs certainly does not indicate any zoological relationship. Most ostracods are small enough to be classed as microfossils, but a few approach the size of a broad bean, and these species can be as conspicuous as brachiopods when they are broken out of the rock. The smaller species, a millimetre or two across, frequently occur in huge numbers, and under the microscope show a beautiful variety of fine detail, which makes them excellent guide fossils (*see* below). The ostracods occur in both marine and fresh water environments (different species of course), and there are specialized species adapted to brackish water conditions as well, so that they are also useful indicators of past sedimentary conditions. The tiny shells are composed of calcium carbonate. The ostracods have a history extending back to the Cambrian, and they continue as a varied and successful group today. They

have been most widely used in dating rocks of Mesozoic age, but there seems every reason to suppose that they will be used more widely in Palaeozoic rocks as well. The electron microscope has given a new dimension to their study, because it enables even the finest details to be seen at high magnification. Although some ostracods, particularly fresh water ones, are smooth and featureless, others are covered with pimples and ridges that make them highly distinctive fossils.

FORAMINIFERANS

We have already mentioned the giants among these single-celled protozoa. Most of the foraminiferans are very small, less than 1 mm in diameter, and must therefore be studied under the microscope. Many of the Tertiary species are found in fairly unconsolidated sediments, and can be extracted from the rocks simply by sieving; the 'forams' can then be picked out by eye under a microscope. The well-preserved fossils are studied using an electron

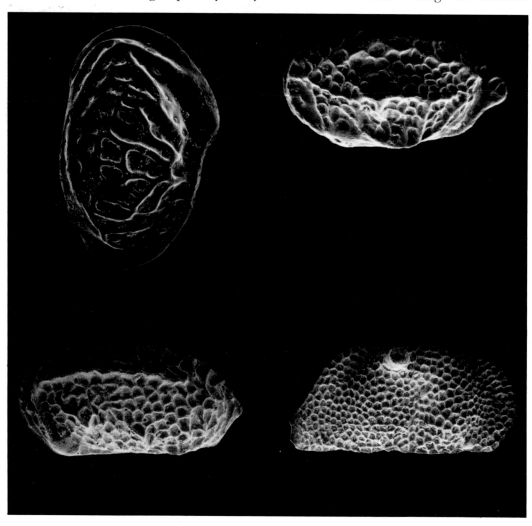

Ostracods: small arthropods with two valves. These examples are Jurassic in age. Magnifications about × 75

Scanning electron microscope pictures of tiny foraminifera: a host of shapes based on a single cell. Magnifications × 100 to × 150

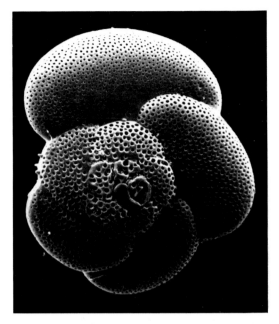

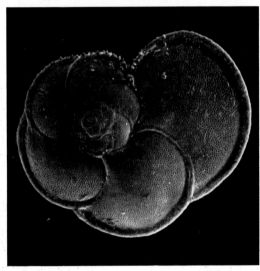

microscope to make perfect, high magnification pictures (micrographs). In spite of their small size many foraminiferans show a wealth of fine detail; some of them are covered with a host of tiny thorns, and many have long spines, or strange, lip-like structures. Their general shape varies from species that look like tiny necklaces, to others that resemble a small bunch of grapes, and yet others with inflated, globular chambers.

They are a tribute to the flexibility of the single cell, and this flexibility means that they are among the most useful of geological clocks for reading out the age of sedimentary rocks. This is especially true of the planktonic species, which are so numerous in the Mesozoic and Tertiary. Oil companies in particular employ numbers of 'foram men' to correlate the rocks in which oil is found, and some of these experts have become renowned authorities on the history of the group. Since drilling on the deep sea floor has become a practical possibility, the foraminiferans have acquired an added importance. Changes in the populations of planktonic species accurately reflect the changes in climate in the more recent geological past, and they can be used to read the ages at which particular pieces of the ocean floor were created at the mid-ocean ridges. Of all the fossil animals it is the tiny foraminiferans which have proved most useful in the revolution of geological ideas that accompanied plate tectonics. They have to be used carefully, though, because some of the shapes that were evolved in the planktonic species were produced independently at different times from different ancestors. Evolution played the same game more than once to produce superficially similar endproducts. Often it is necessary to look *inside* the minute chambers to find out the most intimate details of their construction. Many of the older occurrences are in hard limestones from which it is not possible to extract the entire animals, and here thin sections are the only means of studying the evolution of the group. 'Forams' have been used extensively to correlate rocks of Carboniferous and younger age, and really come into their own in the Tertiary, after the disappearance of the ammonites in the late Cretaceous, where they are perhaps the standard group for the erection of zones. Their ubiquity even led one worker to suppose that *all* rocks (even including igneous ones!) were made of foraminiferans

− an impression which might be forgiven after a hard day sorting out hundreds of the animals from residues.

COCCOLITHS

Even smaller fossils are used to date rocks of Mesozoic and younger ages. Among these, one of the most important are the coccoliths, minute plates a fraction of the size

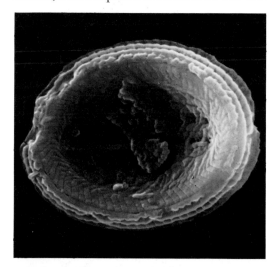

of a foraminiferan (usually with diameters of only a few μm). In 'life' they were the covers for resting cells of single-celled algae (coccolithophorids), with many coccoliths to a single cell. Some of them are so small that they lie at the limit of resolution for a light microscope, and again their study has been revolutionized by the use of the electron microscope. Coccoliths form beautiful rosettes of calcite plates, every species with the plates stacked in a different fashion (*see* below). Although so small, they seem to have changed rapidly through time, and so they are very useful in dating rocks from boreholes and similar small samples. They may even have their use in criminology, as they can be used to identify the source of even the merest smear of sediment on a suspect's shoes. They can form a large part of the fine fraction of sedimentary rocks, as in the soft, white Cretaceous chalk.

MICROSCOPIC PLANT REMAINS

The fossils of spores and pollen are extremely small, but they are surprisingly

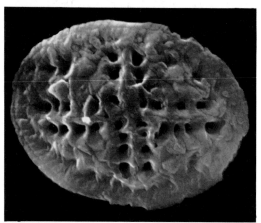

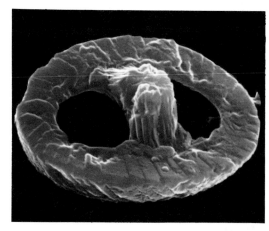

Coccoliths − calcareous platelets of marine algae. The largest is only one hundredth of a millimetre across. Pictures taken with the scanning electron microscope

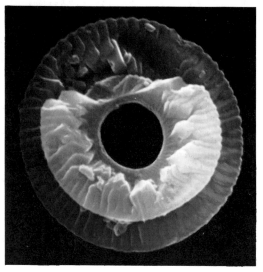

tough. The walls of the tiny objects are composed of a very resistant organic material, that serves in life to contain the vital reproductive material, and has a very high chance of becoming fossilized. So insoluble are the walls of pollen grains that they even resist attack by hydrofluoric acid, possibly the most unpleasant and voracious acid there is. The tiny fossils survive after the rocks that contain them are broken down by acids and other chemicals and it is not uncommon to find these microfossils in rocks that otherwise lack all trace of organic material. This is because pollen grains are wind-borne, and may come to their final rest in sediments from environments in which other life is lacking. Although so small, different spores and pollens have many kinds of peculiarities which allow the recognition of different species under the microscope: they may be variously ornamented with ridges or spikes, inflated or compressed, and have many different shapes (*see* below). So they are widely used in dating rocks, and are of pre-eminent importance in the dating of fresh-water or terrestrial sediments in which the kind of guide fossils, which are typical of marine

Fossil spores: their varied patterns of protuberances make them useful in dating sedimentary rocks.
Bottom: two Devonian acritarchs. Magnification × 1500 approx.

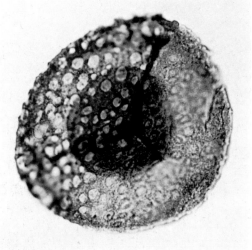

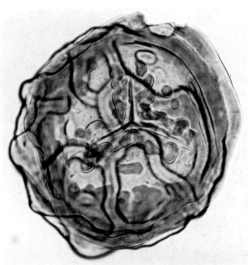

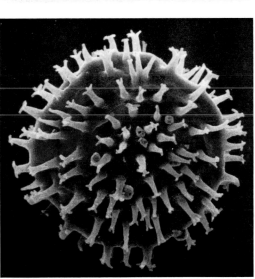

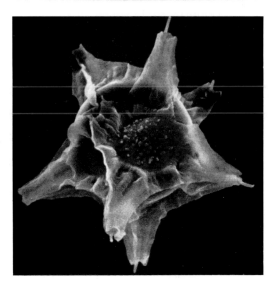

sediments, are absent. So important has the study of spores and pollen become that it even has its own scientific discipline – *palynology* – and palynologists comprise a significant fraction of the palaeontologists employed by oil and mining companies, and geological surveys. The identification of spores is vital to the understanding of both the formation and dating of coal deposits. Because the kinds of plants that produce spores and pollen are Devonian and younger in age, the use of these fossils is confined to the later Palaeozoic, Mesozoic and Tertiary. A particular use for pollen types is in the dating of climatic fluctuations during the Pleistocene Ice Ages; the suite of pollen varieties in these relatively recent rocks can be identified with species still living, so that cold phases will be associated with a predominance of arctic species, and the warm interglacial periods will be reflected in the appearance of the pollen of subtropical species of plants.

Somewhat similar spiky balls can be recovered from marine sediments also. These are the cases of resting cells of marine algae (hystrichospheres). They too have become the subject of wide attention, and palynologists can now contribute much to the dating of marine rocks. Often these tiny fossils can be recovered by treating shales which have no other fossils. A related group of fossils (acritarchs p. 161) can be recovered from the earlier Palaeozoic rocks as well, and even far back into the Precambrian, as we noted in the last chapter. The future of industrial palaeontology will probably be closely linked with these unassuming, tiny balls, which can be recovered from the rocks in their thousands, and can solve the problems of dating rocks from which other fossils cannot be recovered.

COAL AND OIL

In 1769 James Watt patented the design of his steam engine, and so began the plunder of the fossil fuels. The use of coal as a cheap, accessible source of energy fuelled the growth of industry, and gave the ascendancy to the industrial nations in world economics which lasts to the present day. The world of trade can be viewed as a vast machine, one which turns on the consumption of energy mined from the geological record. In this century oil became an even more important source of the energy needed by the machine, which was growing inexorably as population increased, and the material expectations of the workers grew with it. As the complexity and extent of the industrial process has expanded to its present gross proportions a few people have started to question whether the whole machine might grind to a halt. There *will* have to be an end to the reserves of coal and oil, they cannot grow again in the rocks once they have been removed. We have to hope that man's ingenuity will be equal to the challenge of finding replacements for the fossil fuels. Radioactive minerals and solar energy may go some way towards filling the gap, but new sources of energy will also have to be discovered.

When we burn coal or oil we are releasing the energy of the sun beamed upon the earth for hundreds of millions of years. All the energy stored in the ground depended on the photosynthetic activity of plants, whether huge trees in the Carboniferous period or minute, planktonic algae. The formation of coal is a familiar story: it is the black, carbonized, compressed concentrate of plants, especially trees. Coal could only form under certain conditions; it is not sufficient to have a stand of trees that gets swamped by sediment. It takes generations of trees to produce a workable seam of coal. Under normal conditions wood decays, its organic material consumed by bacteria and fungi, until it is friable and porous, and ready to fall back into the soil from whence it came. For trees to turn into coal they have to be protected from normal, aerobic decay. The right conditions for this to happen pertain especially in swampy environments, where large areas may be fetid with lack of oxygen a few centimetres below the surface. In humid, warm forests like many of those of the Carboniferous, plant growth would have been rapid, but even so it would take perhaps several hundred thousand years to accumulate enough material to form a coal seam. It is also essential for the whole area in which the coal was forming to be slowly subsiding, so that in time the tree trunks became part of a thick sequence of rocks. Sometimes, the sea made one of its periodic advances across the coal swamp, causing a retreat of the trees and burying the potential coal beneath a blanket of marine sediment. At other times increased floods of sediment from rivers dumped their sands and silts. As burial of the coal proceeded volatile materials were driven off: it requires a great deal of burial (3000 m or so) before the process is advanced enough to produce a coal of much utility. Very deep burial, or the heating effects of nearby igneous activity, are necessary to produce the highest quality, nutty coal known as anthracite. All this takes a lot of time, and most of the higher quality coals are correspondingly of Palaeozoic age. Mesozoic or Tertiary coals are often known as *lignites* – fossil peats in which the process of coalification has not proceeded to the full degree. Carboniferous coals accumulated in a number of separate basins, and to correlate the rocks between these various basins the spores of the coal measure plants are exceedingly useful. Some coals are almost entirely composed of masses of such spores. The subsequent history of the coal basins was a complicated one. They were nearly always fractured by faults, breaking the coal seams up, and so contributing to the dangers of mining which are part of the folklore of industrial societies. All coals reveal their plant origin in the impressions of bits of bark, or occasionally leaves, that can be seen on the shiny surfaces.

Oil, on the other hand, seems to have no obvious connection with past life. Yet the dark fluid that comes gushing from deep wells to stock the refineries of the world is just as much a product of biological activity as is coal. The ultimate origin of oil is the organic material contained in many sedimentary rocks, including marine mudstones and shales, especially those that accumulated under somewhat stagnant conditions. The fixing of the sun's energy in the sea comes from the action of photosynthetic plants – particularly single-celled algal plankton. This is the ultimate foundation of the whole economy of the sea, and so the 'raw material' for oil is in a sense the produce of fossil sunshine in the same way as coal. Oilfields have been found in rocks as old as Cambrian. The hydrocarbon compounds typical of oil are released from organic material in sediments by bacterial activity, but at an early stage are dispersed throughout the rock. Time and burial are essential to produce a workable oil field. Burial alone probably alters the kinds of hydrocarbons that are present. Eventually during compaction oil is squeezed out, together with pore-water, and then can migrate to sites where it becomes concentrated. These are the *reservoir* rocks, often rocks which would initially have contained no oil at all, but serve to gather it together because they are highly porous, concentrating all the dispersed hydrocarbons from the rocks below into an economic pool. Finally it is necessary to *stop* the migration of the oil, to prevent it from moving all the way to the surface. When this does happen it produces natural 'seeps' of oil, the most famous of which are asphalt 'lakes' like those in Trinidad, and it is probable that many of the first wells were drilled in the vicinity of such obvious surface 'shows'. The search for oil has led to wider and wider exploration. At first the most accessible fields, like those of Texas, were exploited, and were the foundation of many personal fortunes. The vast resources of the Middle East, which are the main source of supply at the present, were tapped from somewhat less hospitable territory, but now any area with sedimentary rocks is explored for its oil potential, whether high in the Arctic, or under the sea. The latter areas are naturally confined to the continental shelf, where the sedimentary cover continues to the continental slope – there is no oil in the deep sea. Some of these new areas have proved highly productive, like the British North Sea field, but sooner or later all the possible sites will have been explored, and then there will be nowhere else to drill. The coal industry may anticipate a renaissance before the end of the twenty-first century.

The classic oil trap is an anticlinal fold in the porous rock, capped by some impervious strata that make it impossible for the oil to migrate further and escape (*see* p. 164). The oil is often capped by a field of gas under high pressure. This is why when the drill penetrates the petroleum field the oil gushes out, and if the gas ignites this can result in spectacular, but dangerous 'blows'. Many other kinds of structures can produce important oil fields. The oil can be concentrated into porous, sandy lenticles within otherwise more shaly rocks (*see* p. 164). In some cases the porous rocks can be fossil reefs, or corals or bryozoans. Reef rocks of this kind usually have a much higher porosity than average because of the gaps between all the frame-building organisms, and since they usually also have a considerable lateral extent, reef rocks at depth and sealed off by less permeable strata can be very important sources of oil. This explains why hard-nosed oil executives support palaeontological research into the geology of fossil reefs. In the Middle East and elsewhere oil traps have formed against the sides of salt domes. These domes of deeply buried salt are actively rising, almost as if the salt were behaving like an igneous magma. The analogy is not altogether misplaced, because the salt does act in the manner of an igneous intrusion, flowing into the domes and pushing upwards under the influence of the load of sediments pushing down on the soft salt deposits on either side (*see* right). As well as sandstones and reef rocks, the magnesian calcite rocks known as *dolomites* are often full of little cavities lined with crystalline dolomite, which may get filled with oil under the right geological circumstances. This is one rock on which the petroleum geologists and the palaeontologist might be forgiven in parting company, because dolomites of this kind are one of the poorest kinds of rocks in which to find fossils. All petroleum-bearing rocks, when outcropping at the surface, have a characteristic smell if freshly broken – rather an unpleasant one, reminiscent of greasy rags discovered in a forgotten corner of the garage.

Nowadays, oil is recovered from greater and greater depths, and it is usually impossible to infer whether the right kind of geological structure is present there just by looking at the surface geology. A great deal of exploration is done using geophysical equipment which senses out the most important structures, and records the differences between permeable rocks like sandstones and impermeable shales and mudstones. Drilling now is nothing like the 'wildcat' operation it once was; though even today luck plays an important part in the productivity of any strike. Once drilling starts the palaeontologist (usually a micro-palaeontologist) comes into his own, identifying the fossils recovered and keeping a log of the age of the rocks through which the drill is passing. In some cases oil is found so consistently in rocks of a particular age it almost looks as if it were seeking out the characteristic fossil. In this circumstance it is likely that the fossil indicates a suitable facies (p. 38) for the formation of an oil reservoir, and here the fossil is used as something more significant than a mere calibration on the age of the borehole.

MINERAL DEPOSITS

Many mineral deposits are associated with igneous and metamorphic rocks, and fossils are scarcely relevant to their understanding. Other minerals are often found in sedimentary rocks, and fossils are used to

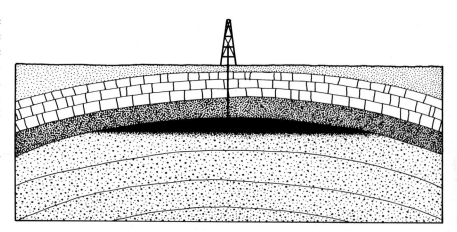

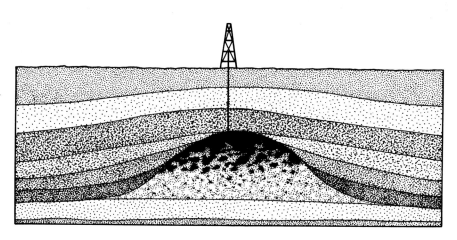

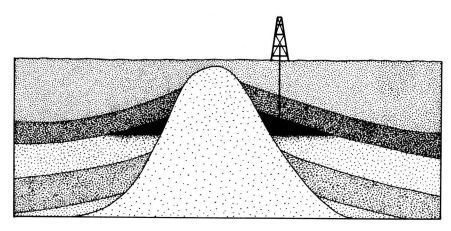

correlate the formations containing such minerals in just the same way as in the oil industry. Some of the commercial deposits of iron are in the form of beds of iron-bearing oolites (p. 41) extending over many square miles. These include normal marine deposits with a characteristic fossil fauna consistently associated with the ores. Phosphate deposits, the basis of the fertilizer industry, are also extensively developed in marine sediments. Often such phosphates are found in sites which were once near the

Oil traps. The different geological circumstances in which one may expect to find accumulations of oil. *Top,* an anticline; *middle,* reservoir in fossil reef; *bottom,* a salt dome trap

The Cotham 'Marble' – an ornamental limestone produced by algae

edge of a former continent. The phosphate was probably introduced into the sediment as a result of the upwelling of deep oceanic water, which tends to occur in such positions, as it does off Peru today. The same situation also prompts a wide variety of animal life – and rich fossil faunas in the past. Phosphatic deposits as old as Cambrian have been found associated with a wealth of trilobites. Rarely the fossils themselves may make up the phosphatic ore, if there is a sufficient concentration of phosphatic shells. Other peculiar 'fossil beds' have special commercial applications. For example, the diatomaceous 'earth' of the West Indies, a rock almost entirely composed of the tiny fossil shells of diatoms, has a variety of industrial uses, including use as a special filter.

ORNAMENTAL STONES

A walk around a European capital city can often be a synoptic guide to the different fossils to be found in that country, or at least to a range of those to be found in limestones. Particularly on older buildings, when builders had time and money, and labour was cheap, the floors, walls, columns and even the ceilings can be covered with a skin of ornamental stone. Many such facings contain sections of fossils. Many of course, do not, being slabs polished from igneous rocks, like granite, or metamorphic ones like marble. But a great variety of limestones have been used as 'marbles' (a confusion of terminology has led all predominantly calcium carbonate facing stones to be called marble, where the geologist restricts the term to metamorphosed carbonate rocks) for ornamental purposes, and often the pres-

ence of fossils lends them their peculiar charm. Coralline rocks are particularly in demand, and the polished sections afford an excellent way of studying the internal structures of these animals. Other common facing stones include those largely made of the stems of crinoids, which make a bold patchwork of rods and struts in white calcite, contrasting with the darker matrix. In Scandinavia there are beautiful red limestones of Ordovician age containing the remains of straight nautiloids. In southern Europe such ancient limestones are not available, but Cretaceous rocks enclosing the peculiar bivalve molluscs known as rudists (p. 47) are often polished to great effect, and even younger Tertiary limestones containing masses of the giant, single-celled foraminiferan *Nummulites* have also been widely used for ornamental purposes, looking like great masses of spiral nebulae adrift in a finely grained groundmass. Limestones formed by layers of algae have been used in the manufacture of ornaments. One such, which was popular in the nineteenth century, is the Triassic Cotham 'Marble' (above) in which there may be imagined a landscape of trees, picked out by the layers of sediment. The Purbeck 'Marble' was extensively used as a facing stone in English cathedrals; columns of this rock surround the main pillars of regular limestone in Salisbury Cathedral. This Jurassic limestone is almost entirely made of the shells of one species of gastropod, and the polished surfaces provide sections through this fossil from every angle. The list could be extended indefinitely: wherever good, homogeneous fossil-bearing limestone is to be found it may be used as a facing stone.

DECORATIVE USE OF FOSSILS

Fossils have been used to manufacture many different kinds of decorative objects. They have been found as talismans associated with cave cultures of *Homo sapiens*. The North American Indians used the small oval trilobite *Elrathia kingi* to manufacture a necklace composed of many examples of this fossil. The same species is mined commercially today to produce everything from tie pins to paperweights. The most consistently used fossil material is amber, fossil resin, which was discussed earlier in this book. Amber fossils include a wide range of insects and spiders. Amber has been shaped into drop-like pieces, and carefully matched for colour, before being mounted in necklaces, pendants, and earrings. Some of the nineteenth century examples of amber jewellery are particularly fine. Fossil wood which has been replaced by the hard mineral silica often retains the finest details of its cellular structure. It takes a very high polish, and is often an attractive deep reddish brown colour when cut. This has been exploited to advantage in the manufacture of a variety of table ornaments, and the larger trunks have been cut through to make an extremely heavy table top, which would certainly be immune to the normal stains of domestic use, and must be the oldest table in existence. In the Jurassic rocks of Yorkshire another kind of fossil wood is preserved as *jet* – a dense, very dark material that is the origin of the phrase 'jet black'. In the nineteenth century jet had a considerable vogue in necklaces and the like (*see* above right). The brilliant black colour of polished jet was adopted by Queen Victoria after the death of Prince Albert, and thence by society, as an ornament that was both decorative and decorous.

Jet ornaments are polished pieces of Jurassic fossil wood

It has recently become a fashion to treat well-preserved, large fossils as objects of beauty in their own right. Fine examples of ammonites or fossil fish can be found on sale at inflated prices, described as 'Nature's Sculpture' and the like. Good looking fossils can now command high prices at auction, and several concerns have grown up in the United States devoted to the commercial exploitation of the biological past. The increased appreciation of the value and beauty of fossils is something to be welcomed. The only problem is that fossils, unlike living animals, cannot breed and replace themselves. Like coal and oil they are a finite resource. Classic fossil localities are easily worked out, and every palaeontologist has had the experience of arriving at a well-known locality to find nothing left but a large hole. Fossils *are* valuable, each one in its way is a small miracle. But they are most valuable for what they can tell us about the past.

CHAPTER NINE

DISCOVERY OF A NEW DINOSAUR – THE STORY OF 'CLAWS'

Occasionally a new fossil discovery is made which is as spectacular as it is important. A find of this calibre was made in January 1983 by William Walker at the claypit of the Ockley Brick Company in Surrey. Mr Walker had a long-standing interest in collecting fossils, and often used to visit localities in southern England after his week's work as a plumber. He was knowledgeable enough to recognize an unusual find, and wise enough to bring it quickly to the attention of the professional palaeontologists. On this occasion he broke open a nodule that was lying in the clay, and recognized that it contained fragments of bone. At home he assembled the fragments and discovered that they made a huge claw – but with the tip missing. He was even able to return to the claypit several days later and find the missing tip encased in the rest of the nodule where it had fallen to the ground. So it was that a major new dinosaur discovery was made – the first remains of *Baryonyx*, popularly known as 'Claws'. No better example could be found to illustrate the progress and problems of palaeontological research.

Mr Walker's son-in-law brought the claw to The Natural History Museum in London, where its unusual features were quickly appreciated. The first concern was whether there might be more of the dinosaur skeleton preserved in the claypit. Fortunately, the clay digging at Ockley is suspended during the winter months so that any other bones still had a good chance of being there if they had not already been incorporated into bricks. In February palaeontologists from The Natural History Museum visited the site, and established that there were indeed more pieces of dinosaur skeleton close to where the original claw bone had been found. Would these prove to be part of the same animal, or would they belong to one of the commoner species? Unfortunately the conditions were so wet and unpleasant that curiosity had to be suspended until later in the year when some good spring weather had dried out the clay sufficiently for serious excavation. Then a team of eight settled on the site and collected every bone fragment they could find; this took three weeks of hard work.

The process of excavation was laborious. Nothing would have been more disastrous than simply picking up the bones that were lying around, and hoping for the best in making a later reconstruction. Instead, every visible piece of bone was located and a certain amount of on-site cleaning was performed. It soon became clear that much of the bone was preserved, like the original claw, inside nodules of a siltstone much harder than the surrounding clay. These bones were going to require careful preparation in the laboratory, only a fraction of the skeletal material being visible on the surface. The bones were distributed in a somewhat higgledy-piggledy fashion, but from their size and arrangement they seemed very likely to belong to one animal – and this was surely the claw-bearing dinosaur. The find was the most exciting kind, a more or less complete animal, even

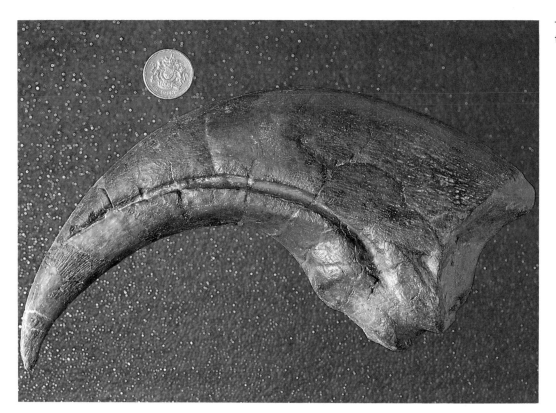

The claw of *Baryonyx*, the first find of a new dinosaur

though its skeleton had broken up in a way which would require patient reconstruction.

The skeleton was preserved in a kind of mosaic, partly exposed and partly concealed inside the siltstone nodular layer. A careful map had to be made to show the exact relationships of each piece – this might prove of vital importance in piecing together enigmatic pieces of bone at a later stage. At least those bones preserved inside nodules were protected from abrasion – but the other bones on the surface required protection if they were to be moved safely to The Natural History Museum. This was done by wrapping exposed bone in tissue and then encasing the whole in a jacket of plaster-of-Paris bandages, much as one might set a broken leg. The mosaic pavement including the dinosaur bones was separated into blocks along natural cracks, and then these could be transported. Altogether some 54 blocks weighing more than two tonnes contained the retrievable remains of 'Claws', a surprising haul when one remembers that Mr Walker was able to hold his original find in two hands.

All this material is now in the Palaeontology Laboratory of The Natural History Museum. Although the discovery was made seven years ago, as I write the excavation of the bones is still in progress. It has proved an extremely laborious process. The bones that were originally partially exposed at Ockley could be prepared relatively quickly. Half the casing of plaster was removed, the other half being left as a support, and then the clay and other matrix was removed by mechanical means – there are a number of special tools to help with this. The exposed bone was then washed and dried, and strengthened using a resin solution. But the bones encased within the hard siltstone have proved much more difficult to extract. The enclosing rock is tougher than the precious bone material. There is really no alternative but to dig out the bone manually by removing the matrix. This is a very skilled job, because one slip may damage a piece of evidence which can never be recovered. Even though a dinosaur was a large animal parts of the skeleton, particularly in the skull, were surprisingly delicate. Hence much of the manual preparation is actually performed under a binocular microscope, so that the appearance of the merest flake of bone will not escape the attention of the preparator. A lot of this skilled work has been undertaken by one man, Ron Croucher, who has had great experience in this kind of preparation. Ron hopes that by the time he retires 'Claws' will be a fully mounted skeleton. Patience is the most important quality of the dedicated preparator. None the less, the assistance of a variety of tools makes the process less laborious than it might otherwise be. A particularly effective tool has proved to be a machine that fires abrasive powder from a compressed air 'gun', eroding the rock away in a speeded up version of natural weathering. It is necessary to protect the bone surface (with a special latex) as it appears, lest it too becomes eroded. Particle by particle the new dinosaur is being exposed to scrutiny.

It soon became apparent that the discovery was even more exceptional than there was any reason to hope. Inside the hard siltstone the bone was well-preserved. There was no way of knowing what the hard nodules contained until the bone had been carefully dug out. One of the most exciting discoveries was a good part of the skull. It is not surprising that the robust limb bones or vertebrae (or even claws) of dinosaurs are the most durable fossils, being both large and tough. The skull, apart from the teeth, is actually more delicate. It is also of special interest to the scientist in determining the evolutionary relationships of the dinosaur, not to mention those interested in making a dramatic reconstruction of the whole animal. So the discovery of the skull of 'Claws' drew dinosaur palaeontologists from around the world to speculate about what kind of animal it was. Once these important parts of the skeleton had been discovered and prepared, casts of them could be made to send to other scientists who were not able to travel to London. In such a way is scientific information disseminated through the international community.

However, by early 1986, when all this new information was available 'Claws' still did not yet officially exist. This is because it had yet to be formally christened in a

Baryonyx in the process of being excavated from the Ockley Brick Pit

scientific journal, as was explained in Chapter 4. 'Claws' had already been reported in newspapers in almost every European country ('Claw Blimey! A Dinosaur' as one tabloid put it) but this reportage did not count in the scientific world. To make good this omission, my colleagues Alan Charig and Angela Milner provided a preliminary description of the animal, based on what was known up until that time, which was eventually published in the journal *Nature* in November 1986. At the moment of publication 'Claws' was named *Baryonyx walkeri*, and this will be the name by which the dinosaur will be known in perpetuity. In a few years *Baryonyx* will join *Tyrannosaurus* in the small boy's dinosaur demonology, but in the mean time Ron continues his slow and meticulous preparation so that we may know as much as possible about the whole animal. The generic name *Baryonyx* preserves the idea of

'Claws' because it is Greek for heavy talon or claw. The species name *walkeri*, of course, commemorates William Walker's discovery of the original claw – and to be honoured in this way is a just reward for his perspicacity. Charig and Milner decided that the new dinosaur was distinct enough to be placed in a new family as well, which is some measure of its scientific interest.

It is now clear that enough bones of *Baryonyx* are known to prepare a reasonably confident reconstruction. However, it will be recalled that mistakes have been made in reconstructing dinosaurs in the past, and so it is wise to exercise a certain caution. Sadly, much of the tail has proved to be missing, but this is the least important part of the skeleton because the tails of dinosaurs have much in common, and that of *Baryonyx* can be reconstructed from the few fragments to hand by comparison with other species. Likewise missing ribs are not a serious

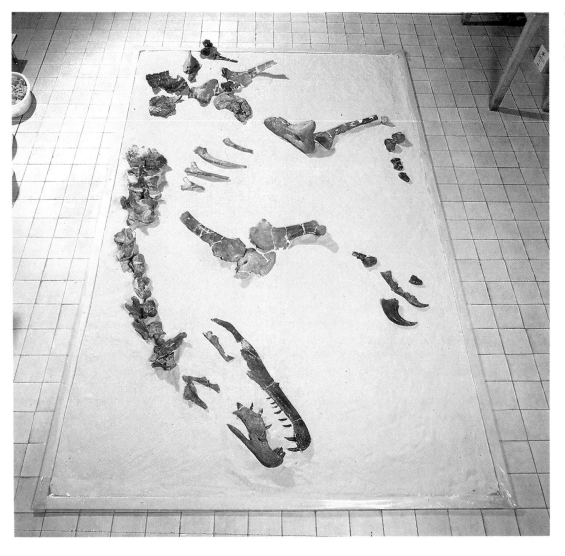

The present state of knowledge of the skeleton of *Baryonyx*. With each new discovery more of the skeleton is being completed

deficiency.

When the claw of *Baryonyx* was discovered a first comparison was suggested with a small Cretaceous dinosaur from Montana called *Deinonychus*. This remarkable animal had the second toes on its hind legs extended into enormous claws (its name means 'terrible claw') which could be kicked into action as dangerous weapons. Could the new claw have come from a comparable animal? *Deinonychus* had not been recorded outside the USA. As more of the bones of *Baryonyx* were prepared it became clear that it was an entirely different animal from *Deinonychus*: to begin with, the claw was on the front limb. However, it certainly belonged to the group of meat-eating dinosaurs that included not only *Deinonychus*, but *Allosaurus* and *Tyrannosaurus* as well. This group is known as the theropods. This did not diminish the excitement of the discovery, because the fossil record of theropods of Lower Cretaceous age was very poor and *Baryonyx* was a major new addition to our knowledge. Furthermore it was only the second reasonably complete theropod to have been found in Britain, and the previous record was discovered more than a century earlier. It may well be another century before the Weald yields the next theropod.

It is now clear that the skeleton of *Baryonyx* has some curious features. Overall it is somewhat less massive than *Tyrannosaurus rex*, although it may have been as much as 10 metres long. Its front limbs are surprisingly well-developed; even if it walked on its hind limbs it may have been able to move on all four legs, something which no other theropod did. *Tyrannosaurus* had the front limbs reduced to tiny remnants. The *Baryonyx* skull was rather narrow, and prolonged into a shape not unlike a crocodile's snout, and, again like the crocodile, its jaw line was twisted into a crooked sneer. *Baryonyx* had about twice as many teeth as other flesh-eating dinosaurs. Its nostrils were rather far back. The famous claw is believed to have comprised the thumb of the front limb. Like the talon of an eagle the claw bone would have been covered with a horny layer – making it perhaps 35 cm long.

Now that we know these facts it is possible to develop a theory about how *Baryonyx* lived. Drs Milner and Charig believe that *Baryonyx* was a fish-eating dinosaur living on the edges of the Wealden lakes. The long snout with numerous teeth would have grasped the fish effectively, and then the dinosaur would probably have swallowed it whole, head first, as do living fish-eating crocodiles and herons. This is in great contrast to its relative *Allosaurus*, which had a comparatively massive head, and teeth arrayed to help the shearing of flesh. The remains of a large and partially digested fish, *Lepidotes*, have been found in association with the *Baryonyx* skeleton; this metre long fish would make a substantial meal even for a dinosaur that may have weighed two tonnes. No doubt if the opportunity arose *Baryonyx* would not have been reluctant to scavenge, particularly if the bloated body of the plant-eating dinosaur *Iguanodon* had drifted into its lake

Possible candidates for the relatives of *Baryonyx*: *Allosaurus* and *Deinonychus*. Subsequent work showed that *Baryonyx* was not closely related to *Deinonychus* despite their both having 'claws'

habitat, which would have provided rich pickings.

Iguanodon is the commonest dinosaur in the same quarry as yielded *Baryonyx*, and it was doubtless commoner than *Baryonyx* in the Lower Cretaceous also. It probably fed on the lush vegetation that grew around the Wealden lakes. From the associated fossils and geology quite a lot is known about the environment in which *Baryonyx* fished. The Weald of southern England was part of a great area of warm, low lying and marshy territory, with numerous rivers and lakes, populated by turtles and crocodiles. Horsetails and ferns flourished in the swampy ground, and the humid atmosphere teemed with insects. In such a setting, it is not stretching the imagination to see *Baryonyx* lurking in the swampy vegetation waiting the moment to grasp a sluggish and unwary fish in its crocodile-like jaws. On death, the body of *Baryonyx* drifted into one of the lakes and sank, to be covered with silt – which eventually became the siltstone that contained the fossils. It is perhaps not very likely that our *Baryonyx* specimen was killed by a large predatory dinosaur (the fierce *Megalosaurus* is also known to have been present in the Cretaceous of the Weald) because of the comparative completeness of the skeleton. The conditions favoured by *Baryonyx* may have existed over a very wide area. More than 20 years ago some fragmentary dinosaur remains were discovered in the Sahara Desert in Niger by French palaeontologists. Although the deposits containing them were a little younger than those which yielded *Baryonyx*, a claw and a snout fragment were so similar to those of the Wealden dinosaur that there seems little doubt that a close relative lived in this part of Africa at a time when conditions were much more favourable there than they are now. An expedition sponsored by The Natural History Museum went to Niger in 1988 to try and find more of this animal. Sadly no bones of this interesting dinosaur were discovered, although the remains of several other kinds of dinosaurs – some of

them not hitherto known from Africa – provided an adequate compensation.

This brings us back to the claw that first attracted Mr Walker's attention in the Ockley pit. It should be possible to fit this formidable appliance into the picture of *Baryonyx* life habits. There is nothing quite like it among living animals, so its function is difficult to establish. If the interpretation of *Baryonyx* as a fish-eating dinosaur is correct we might imagine the claws as an important part of this feeding habit. For example, could *Baryonyx* have hooked out its prey by a swift dash of the forelimbs before grasping it in its jaws? One thinks of the way grizzly bears stun and hook out running salmon from rivers using their strong paws – and their claws. This explanation would also be consistent with the comparatively long and powerful forelimbs of *Baryonyx* compared with its relatives. After all, it would have been impossible for *Tyrannosaurus* to have used its shrunken forearms in this fashion. The claws of 'Claws' seem rather readily explicable in this way, which squares with all the facts. But one of the charms of palaeontology is that the discovery of one new fact can overturn the most convenient explanation. Like Mr Walker's dinosaur, the unexpected turns up from time to time. It is unlikely that we have found out all there is to know about *Baryonyx*.

Reconstruction of *Baryonyx* as it might have looked in life in its natural habitat

CHAPTER TEN

MAKING A COLLECTION

Making collections seems to be almost an instinct with many people. A fossil collection is one of the easiest to store and maintain, and is the best way to get to know the many kinds of organisms in their various modes of preservation. Some people may prefer to make collections of fossils from the rocks in the vicinity of their own home, other may deliberately try to get a wide coverage of the different kinds of fossils from rocks of diverse ages. It is surprising how quickly a collection grows; if given a chance it soon begins to oust the collector from house and home. This chapter gives a little practical advice on how to find fossils, clean them, identify them and store them. There is no doubt that some people have an almost supernatural faculty for finding fossils. Fossils seem to fall out of the rock for them, whereas others labour long and hard with no reward. But in general, persistence pays. There is no substitute for returning time and again to the same site, really pounding the rocks, and carefully examining anything that looks organic. The only two essential tools for this are a good geological hammer, and a hand lens. Geological hammers come in two main varieties: those with a wooden haft, and those in which the head is welded to a hardened steel haft. Either is perfectly satisfactory for hammering sedimentary rocks. Otherwise all that is needed are old newspapers (for wrapping) and patience.

WHERE TO FIND FOSSILS

There are no special rules about where to find fossils. Obviously there are certain restrictions: if rocks have been heavily metamorphosed there is a slim chance of recovering good fossils, but even this is not invariable, and there are examples known where fossils seem to have survived the most appalling maltreatment of heat and pressure. In general, though, good sedimentary rocks are fossil-bearing. There are a large number of known localities, that is, places where fossils have been recovered for a long time, and which are recorded in the geological and popular literature. At such places fossils are sure to be found, but it has to be remembered that generations of hunters have been there before, and that the inspiring specimens now residing in museums were probably collected a hundred years before, when the ground was in prime condition. Nowadays the best collections can often be made from the same geological *horizon* as the classic localities, but from sites a few hundred metres to a kilometre or two away. Then there are some whole areas within which the rocks are thoroughly saturated with fossils; in these cases it is just a question of having enough time to gather all that may be collectable. The Jurassic rocks of the Dorset coast in southern England are one of the best known rock groups of this kind. Formation succeeds formation, all of them rich in fossils. Even if the most spectacular fossils have been removed there is no chance of the localities being 'collected out'. The same might be said of Cretaceous deposits in large areas of the interior of the United States, or the Devonian of Canada, and so on. Such areas are certainly the most encouraging in which to make a first collection. They usually represent former deposition in shallow marine habitats, rich in life, where the sediments have been hardly disturbed subsequently, and where relatively soft sedimentary rocks yield up their treasures with little resistance.

Such prolific sites apart, most sedimentary rocks can be made to yield fossils with prolonged searching. Sandstones in general are the least rewarding. Some sandstones were deposited in desert environments and scarcity of fossils in these is hardly surprising. Others are turbidites (p. 43) in origin, and fossils are rare in these too, although the presence of trace fossils should not be ignored. Limestones are only occasionally lacking any organic remains (apart from Precambrian ones). If the reader is attracted by finding older fossils there is the increased possibility of structural complication in sedimentary rocks of this antiquity. Cambrian to Devonian rocks (and younger rocks too in some areas) are often affected by a cleavage, which makes the rocks split in a direction unrelated to the original bedding, in which the fossils lie. Even so, it is quite feasible to recover the fossils by looking for the traces of the true bedding, and smashing the rock so that it breaks in approximately the right direction, and, with luck, around any fossils. This can be a heart-breaking process, trying the patience of even a professional. If the rocks are folded as well as cleaved, it is possible to find places in which the cleavage and bedding coincide, and this is the premium site in which to search for fossils.

Many rock formations, although they do yield fossils, have them concentrated into small pockets in certain localities. This particularly applies to formations of freshwater or terrestrial origin, probably reflecting the patchy distribution of lakes and pools where fossils can be preserved. The Old Red Sandstone formations of Wales and Scotland locally contain fish, eurypterids and other exciting fossils, although great stretches of the rocks can be disheartening to the casual searcher; here persistence, and a little luck, are indispensable, but of course there is an extra thrill to the discovery of any well-preserved fossil. Coal measures usually yield fine remains of plants, but the best of these are not in coal itself, but in associated shales, and there is always the possibility of finding one of the seams in which insects, like giant 'dragonflies', or vertebrates are also preserved. Rarer fossils are often concentrated in this way into particular bedding planes, and so are easily worked out by the over-zealous collector. It is essential to note the *exact* horizon of any unusual find, so that it can be located again.

The kinds of exposure in which it is possible to find fossils are legion. The most obvious exposures are those in sea cliffs, and these also provide the best sections through different formations. In desert regions exposure can be as good or better, and the same is true of Arctic areas or high mountains, although collecting there is not without its attendant hazards. Never take chances climbing high or crumbling rocks on the supposition that the best fossils are to be found slightly out of reach! In domestic landscapes like that of England most of the best inland exposures are in quarries or on road cuts. It is always worth examining temporary exposures cut during road widening. It is here that local knowledge is at a great advantage: if you are 'on the spot' you can collect quickly before the site is backfilled and the chance lost for ever. Just occasionally temporary cuts open up seams of important fossils that have never been found before. Finally, it is always necessary to ask permission to enter any working quarry; most quarry owners are quite happy to let in the foraging palaeontologist, but they do not wish to have accidents happening on their property.

In hill regions it often happens that the best sections are along the sides of streams that cut through into the bedrock. The same caution applies to asking permission from farmers before trespassing on to their land; nearly all palaeontologists have stories about encountering the wrong end of a blunderbuss after unwittingly straying on to an irate farmer's property. On very overgrown sites it is an advantage to add a *grubbing mattock* to one's equipment, with which to hack out the weathered rock to get access to a clean face.

Regions well-known for their geology often have guide books published about them, by Government Surveys (like the British Geological Survey in Nottingham) or by local natural history societies and museums. Alternatively, nearly every area has a geological map, published by the National or State Survey, showing the pattern of outcrop of the rocks in the area. Maps show the different rock formations. So by tracing the locality of an outcrop on to the map, it will be possible to identify the geological horizon from which any collection is made. As far as Britain is concerned the British Geological Survey publishes a series of regional guides to geology, giving details both of the main rock formations and of the fossils to be found in them; the Geologists' Association also publishes a very useful series of geological guides and excursions.

It is essential to note exactly where the fossils were recovered, both from the point of view of making a personal record, and as an indispensible aid in their subsequent identification. It often helps to make a sketch of the site (*see* below) with the horizons noted from which the fossils were recovered, preferably in metres from some distinctive 'marker' which can be readily identified on a return visit. The different kinds of rock should also be remarked, together with any peculiarity of particular beds, like the presence of trace fossils, or a change in colour. Such observations are the basis of stratigraphy, and it is amazing how a well-kept notebook serves to jog the memory when the fossils are being examined back home.

Page from a field note book, showing sketch of site, and location of fossils

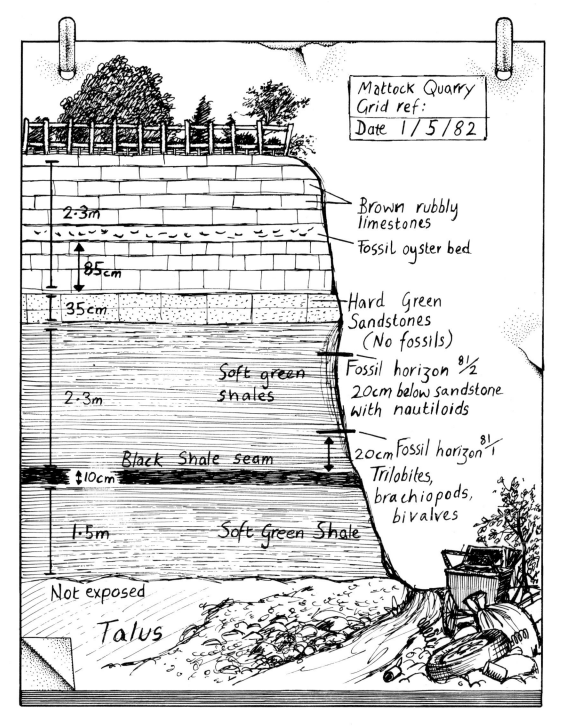

Mattock Quarry
Grid ref:
Date 1/5/82

2·3m — Brown rubbly limestones
— Fossil oyster bed
85cm
35cm — Hard Green Sandstones (No fossils)
Soft green shales — Fossil horizon $8\frac{1}{2}$ 20cm below sandstone with nautiloids
2·3m
20cm Fossil horizon $8\frac{1}{1}$ Trilobites, brachiopods, bivalves
Black Shale seam
↕10cm
1·5m — Soft Green Shale
Not exposed
Talus

CLEANING FOSSILS

There is always a temptation to chip out a sea urchin or a trilobite as completely as possible while still in the field, just to get a proper look at it. This is nearly always a mistake: a careless blow with a hammer or mattock can destroy what might, with patient cleaning, turn out to be a fine specimen. It is far better to wrap the specimen up carefully for cleaning at home. It is important to keep the other half (counterpart) of the specimens, which often show details lacking on the specimen itself. Most fossils are not so fragile as to need special treatment in the field; careful wrapping is sufficient. Some are so delicate that they have to be toughened up before removal, by using solutions of resins.

If one is lucky enough to stumble across the remains of a large vertebrate the best thing is to leave it exactly where it is, and contact a museum with the expertise to extract it properly. Otherwise, it is possible to destroy unwittingly vital information.

Once the fossils are safely home and unwrapped, the cleaning process can begin. Some fossils tend to crack out completely, graptolites in shales for example, and little needs to be done to these. Sometimes, however, the shale still partly covers the specimen, and in this case a sharp tap with a small chisel usually suffices to break the shale along the bedding plane containing the rest of the fossil. The crucial point to remember when trying to dig out a fossil is that damage will result if the cleaning implement is much harder than the fossil and used directly against it. Fossils in shales and mudstones tend to be soft, and the greatest care is necessary to avoid damaging them. A needle mounted in a pin vice can be used gently to squeeze off any covering shale, but this should be done by pressing obliquely on to the covering rock, and not by stabbing at the fossil itself, which nearly always results in unsightly pin pricks.

In some cases the enclosing matrix is softer than the fossil. This is the easiest case to deal with, because the enclosing rocks can usually be removed by scrubbing. Fossils from the Cretaceous chalk can be cleaned using a small bristle (but not wire) brush with water. Cleaning is much more difficult if the matrix is hard, and about the same hardness as the fossil material. There is usually no choice here but to clean off the matrix by hand gradually, relying on the tendency for rock to break off around the fossil rather than through it, at the interface

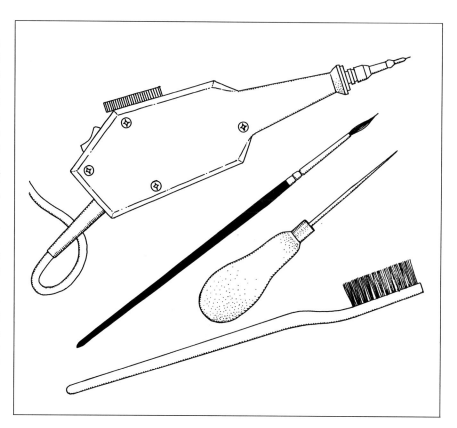

between the fossil and the enclosing rock. A mounted needle can be used for this (the best are the needles used for 78 rpm records, which can still be bought in junk shops), and there are various manufactured appliances that allow the needle to vibrate very fast, producing a percussion that chips away the matrix more rapidly. Cleaning fossils in this way is quite a skilled operation, and if this is being tried for the first time it is best to start on one of the more unimpressive specimens, however tempting it might be to start with the prize exhibit. Sometimes nature lends a hand with the process, and natural weathering already etches out the slightly harder fossil from the surrounding rock. Many of the most impressive museum specimens have been prepared naturally for exhibition in this way, but the majority of these were collected in the last century from virgin sites. These have since been picked bare of such treasures, and nowadays there is usually no substitute for hard work.

In sandstone preservation particularly, but also in some limestones, the actual shell of the fossil has often been dissolved away, and what remains are internal and external moulds. The internal mould usually comes away with little trouble. The external mould should be washed clean of any dirt, which may have to be removed by gently rubbing

Implements for cleaning fossils

with a tooth brush. A perfect replica of the exterior surface of the fossil may now be made by taking a cast from the external mould. Various preparations are suitable for this purpose: a plasticine squeeze can give a quick impression, but does not take up the finest detail. For this a latex rubber solution should be used, or some of the modern resin preparations, the latter having the advantage that the casts are more or less permanent. In fact, casts of this kind are every bit as good, and in some cases better, than having the 'fossil itself', particularly since the internal mould also gives you all the details of the internal surfaces of the fossil.

Special tricks can be tried with limestones. In some limestones the fossils have been silicified (replaced by silica) while the matrix consists of calcium carbonate (p. 17). In this case the rock can be digested in *dilute* acetic or hydrochloric acid (with care!), and the fossils will be left behind. After repeated washing and drying a collection of fossils in this kind of preservation can be mounted in slides, or in small boxes. Any phosphate fossils (such as inarticulate brachiopods) are left untouched when limestones are dissolved in acetic acid, and some beautiful and surprising fossils can be recovered in this way, even when there is not much to show on the surface of the rock.

IDENTIFYING FOSSILS

While it is satisfying to make a collection of beautiful specimens of fossils for their own sake, it becomes even more so if the specimens are identified and classified. Many people find that a general collection of fossils soon begins to take up too much room, and that they prefer to specialize in one kind of fossil (such as trilobites) or in those from a particular area or age. Putting a name to a fossil may seem a rather arid exercise, and so it is by itself. But the real point is that it is not possible to identify correctly a fossil without looking at it very hard, really appreciating the fine points of its construction, and becoming familiar with a whole range of related animals or plants. So to identify is partly to understand.

To identify the general kind (e.g. phylum) of animal or plant is usually easy, as was shown in Chapter 4. With experience it is quite easy to identify the fossil to quite narrow limits, say to order level. There are only a relatively few major kinds of brachiopods, for example, and their general features can be mastered with only a little experience. The problems start when a precise identification is required, to genus or even species. There is no easy way to do this, and sometimes even the expert in the group of fossils will have problems. In some cases a species identification may not even be possible: for example, many brachiopods are identified from their internal structures, so if your fossil does not show these it can scarcely be specifically identified. It is vital to know the precise formation and locality from which the fossils were recovered. Fortunately there are reference collections in museums, which have fine specimens on display, and which have (one hopes) been identified by an expert. These are often arranged rock formation by rock formation, and so it is possible to home in on a series of species with which the one in hand is to be compared. It usually happens that your specimen is not *exactly* the same as the ones on display; the best name is then that of the closest matching species. Most fossil animal populations, like many living ones, include a certain amount of variation, so that it is not likely that any two fossils will be *precisely* the same even if they belong to a single species. Even among palaeontologists there are arguments about whether or not a population of fossils slightly different from other examples of a species is really a different species or just a local variant (see p. 178).

The next best thing to using a comparative collection to identify a fossil is to use books with pictures of fossils as the basis for identification. Any available books only give a selection of typical fossils from a region, but this is a good start, and because most books will figure the most common fossils you are likely to have at least some of your finds illustrated. The Natural History Museum produces three guide books to British fossils, which include drawings of several thousand of the more common fossils to be found in the British Isles. Often, though, a fossil is found which cannot be closely matched to any of the species illustrated in these books. In this case probably the best approach, if the fossil is of an invertebrate, is to obtain the relevant volume of the *Treatise on Invertebrate Palaeontology*. This rather intimidating work is a compilation of all the different kinds of fossils known at the time it was published (different years for different volumes); volumes concern one type of

fossil (e.g. brachiopods or trilobites), and contain descriptions and figures of all the different genera of the animals in question known at the time. It can take a while to learn how to use one of these volumes, and the pictures vary in quality. Of course, this only enables an identification of the genus of the fossil, but this is adequate for most purposes. The volumes of the *Treatise* are not likely to be stored in a local library, but they can be ordered through libraries, and they are certain to be stocked in university libraries. Finally, many museums offer an identification service. The Natural History Museum in London will undertake to identify any fossil that is submitted to it. Occasionally this kind of enquiry will result in the discovery of a new species, and then the finder may be asked to donate it so that it can become subject to proper scientific scrutiny. This will show again how important it is to record the precise details of where a fossil was found. New discoveries are quite often made by amateur collectors, particularly when they return again and again to a favourite site to find the rarer fossils in the fauna. One of the greatest excitements of palaeontology is the possibility that something entirely new will turn up; every hammer blow could be the one that breaks out the new discovery. It is this

that sustains the searcher through long hours where nothing at all is discovered.

STORING A COLLECTION

As the man who made a fortune from selling 'pet rocks' in the United States shrewdly realized, fossil specimens are remarkably easy to look after. Most fossils do not deteriorate with time, and all they require is room for storage in a dry place. A few fossils look superficially attractive if they are varnished: this is not a good practice, however, because the varnish actually obscures a lot of finer detail, and is very difficult to remove once it dries. All fossil specimens should be labelled either with a direct identification and locality, or with some sort of code which refers you to a catalogue. When a collection begins to get large it is impossible to remember all the details about when and where a fossil was collected, and how it was identified. And it is amazing how quickly a collection does begin to get extensive.

Some kinds of fossils are prone to decay. This is particularly true of those preserved in iron pyrites, which includes ammonites and Tertiary fruits and seeds. After a while these can begin to acquire feathery growths of crystals as the pyrite begins to react with

Variation in fossil species: a variable brachiopod. Numbers and development of ribs are different from one specimen to another – but they all belong to a single species the atmosphere. The eventual outcome is that the fossil collapses into a heap of dust. Varnish does not greatly inhibit the process of decay. There are special inert fluids on a silicone base in which fossils like these can be stored. A cheaper alternative is glycerine, but this is messy stuff, and gradually takes up water from the atmosphere, so it has to be stored in air-tight jars.

How the fossils are arranged is a matter of taste. Some people prefer to order them phylum by phylum, others according to their geological age. If the intention is to make a very detailed collection from one area or formation, the arrangement might be by locality. In many ways the last kind of collection is the most satisfying, because it does not take long before the commoner fossils become old friends, and can be used to trace a single geological horizon from one locality to the next. With prolonged collecting, specimens as good as any found in museums will turn up, and the collector himself will begin to know the faunas as well as any expert. Geological time is immensely long, and the volume of fossil-bearing rock passes computation. There is plenty of scope for those with time and patience to make a real contribution to the knowledge of the history of life. The evidence is just waiting to be collected.

POSTSCRIPT

Although there is no shortage of fossils they can be collected only once. Having survived the vicissitudes of millions of years of burial, when they are disinterred they become the responsibility of the finder. Equally, the locality itself is a precious entrance into the past, and is to be respected. There is nothing more depressing than reaching a known locality only to find it mutilated and degraded by some irresponsible collector trying to get the biggest and best specimen without regard for those who follow, the ground littered with broken fossils, which, with a little care, could have been cleaned and saved. In Britain a number of sites have been declared Sites of Special Scientific Interest, which means that they are somewhat protected from over-collecting. The pursuit of the perfect specimen should not be allowed to overshadow the scientific importance of the humbler, fragmentary fossils. The whole point of palaeontology is to reveal the biological history of the world, and fossils are an indispensible means to this end. Any unusual find might prove to be the clue to understanding some new aspect of this history. The evidence of past worlds has to be respected.

GLOSSARY

I have generally tried to avoid the use of technical terms and, where they have been introduced, I have defined them. Some of these essential terms re-appear in the book subsequently without explanation, and it may prove useful to have a short glossary recapping their meaning.

Abyssal At great depth, off the edge of the continental shelf.

Appendages The limbs, gills and antennae of arthropods.

Aragonite Form of calcium carbonate employed in the construction of shells by some marine animals.

Astogeny Growth of a colonial organism by addition of individuals of the colony.

Bedding plane Plane parallel to the former sea floor (or freshwater equivalent).

Benthic (or benthonic) Describing organisms that live (or lived) on the sea bottom.

Calcite The common mineral form of calcium carbonate, of which many fossils are made.

Chert A hard sedimentary rock composed of fine grained silica.

Cleavage Tendency for metamorphic rocks (eg slates) to break at an angle to the bedding plane (hence, plane of cleavage).

Conglomerate Coarse, pebbly sedimentary rock, such as the deposit of an ancient beach.

Correlation The process of establishing the time-equivalence (or otherwise) of sequences of rocks.

Epicontinental Surrounding the edges of continents; especially of shallow seas.

Era Major division of geological time: Proterozoic, Palaeozoic, Mesozoic and Tertiary (Kainozoic, Cainozoic or Cenozoic) and Quaternary.

Exoskeleton The exterior skeleton of the arthropods.

Facies Rock type (or collection of rock types) representing a particular environment where the rocks were deposited (eg reef facies, lagoonal facies); also applied to faunas reflecting the same environment.

Fauna An assemblage of fossil animals from one site or age (a flora is the botanical equivalent).

Filter-feeder Animal that lives by filtering out small particles (usually plankton) from the sea, or fresh water.

Helical coiling The upward-spiral kind of coiling typical of many snails and a few ammonites.

Homeomorph An animal that resembles another, possibly because of a similar mode of life, but is not really biologically related.

Igneous Rocks that have formed from the cooling of hot, liquid *magma*, ultimately derived from deep in the earth, and of course without fossils.

Internal mould Natural cast in sediment of the *inside* of a shell or other fossil.

Intrusion Mass of igneous rock, which intrudes into the surrounding strata.

Living fossils Term applied to animals or plants that have survived for a long time, or at least from a time when there were many more of their kind.

Lower Palaeozoic Term meaning the Cambrian to Silurian periods inclusive.

Marine transgression Invasion of the sea over land area, caused by relative rise in sea level.

Metamorphic Rocks that have been altered by heat and/or pressure, usually because of deep burial or involvement in mountain-building episodes.

Oolite Sedimentary rock (usually limestone) composed of spherical ooliths, and usually formed in shallow water.

Pelagic Free swimming in the oceans.

Phyla (singular, phylum) Major zoological unit of classification, indicating broadly related animal groups.

Planktonic (or planktic) Passively floating in the oceans (noun, plankton).

Radiometric age Age given in years using the natural 'clock' of radio-active minerals.

Regression Draining of the sea from a continental area; the opposite of transgression.

Shield areas Large areas composed of Precambrian rocks that have acted as coherent, stable blocks for hundred of millions of years.

Silica One of the most abundant compounds in nature, silicon dioxide, of which quartz and chalcedony are two of the commonest forms, and which is used as a skeletal material by a few organisms.

Stratigraphy That part of geology concerned with the description of the relationships of rocks in the field, and their correlation.

Subduction zone Area where oceanic crust is plunging downwards beneath an adjacent continental block.

Suture line Line marking the junction of the chamber wall with the shell in ammonites and nautiloids.

Test The 'shells' of echinoderms or foraminiferans

Turbidite Sedimentary rock formed by the action of a turbidity current (*see* p. 43).

Unconformity Break between two sequences of rocks; the lower sequence is often tilted, uplifted and eroded before the deposition of the overlying one (angular unconformity).

Zone Basic unit for correlation, a unit usually typified by a characteristic assemblage of fossils belonging to one or more groups of organisms.

Zooid Individual animal in a colonial animal – applied particularly to bryozoans and graptolites.

INDEX

References in *italics* are to captions, those in **bold** to colour plates.